Ergebnisse der Anatomie und Entwicklungsgeschichte
Advances in Anatomy, Embryology and Cell Biology
Revues d'anatomie et de morphologie expérimentale

Springer-Verlag · Berlin · Heidelberg · New York

This journal publishes reviews and critical articles covering the entire field of normal anatomy (cytology, histology, cyto- and histochemistry, electron microscopy, macroscopy, experimental morphology and embryology and comparative anatomy). Papers dealing with anthropology and clinical morphology will also be accepted with the aim of encouraging co-operation between anatomy and related disciplines.

Papers, which may be in English, French or German, are normally commissioned, but original papers and communications may be submitted and will be considered so long as they deal with a subject comprehensively and meet the requirements of the Ergebnisse.

For speed of publication and breadth of distribution, this journal appears in single issues which can be purchased separately; 6 issues constitute one volume.

It is a fundamental condition that manuscripts submitted should not have been published elsewhere, in this or any other country, and the author must undertake not to publish elsewhere at a later date.

25 copies of each paper are supplied free of charge.

Les résultats publient des sommaires et des articles critiques concernant l'ensemble du domaine de l'anatomie normale (cytologie, histologie, cyto et histochimie, microscopie électronique, macroscopie, morphologie expérimentale, embryologie et anatomie comparée. Seront publiés en outre les articles traitant de l'anthropologie et de la morphologie clinique, en vue d'encourager la collaboration entre l'anatomie et les disciplines voisines.

Seront publiés en priorité les articles expressément demandés nous tiendrons toutefois compte des articles qui nous seront envoyés dans la mesure où ils traitent d'un sujet dans son ensemble et correspondent aux standards des «Résultats». Les publications seront faites en langues anglaise, allemande et française.

Dans l'intérêt d'une publication rapide et d'une large diffusion les travaux publiés paraitront dans des cahiers individuels, diffusés séparément: 6 cahiers forment un volume.

En principe, seuls les manuscrits qui n'ont encore été publiés ni dans le pays d'origine ni à l'étranger peuvent nous être soumis. L'auteur d'engage en outre à ne pas les publier ailleurs ultérieurement.

Les auteurs recevront 25 exemplaires gratuits de leur publication.

Die Ergebnisse dienen der Veröffentlichung zusammenfassender und kritischer Artikel aus dem Gesamtgebiet der normalen Anatomie (Cytologie, Histologie, Cyto- und Histochemie, Elektronenmikroskopie, Makroskopie, experimentelle Morphologie und Embryologie und vergleichende Anatomie). Aufgenommen werden ferner Arbeiten anthropologischen und morphologisch-klinischen Inhaltes, mit dem Ziel die Zusammenarbeit zwischen Anatomie und Nachbardisziplinen zu fördern.

Zur Veröffentlichung gelangen in erster Linie angeforderte Manuskripte, jedoch werden auch eingesandte Arbeiten und Originalmitteilungen berücksichtigt, sofern sie ein Gebiet umfassend abhandeln und den Anforderungen der „Ergebnisse" genügen. Die Veröffentlichungen erfolgen in englischer, deutscher oder französischer Sprache.

Die Arbeiten erscheinen im Interesse einer raschen Veröffentlichung und einer weiten Verbreitung als einzeln berechnete Hefte; je 6 Hefte bilden einen Band.

Grundsätzlich dürfen nur Manuskripte eingesandt werden, die vorher weder im Inland noch im Ausland veröffentlicht worden sind. Der Autor verpflichtet sich, sie auch nachträglich nicht an anderen Stellen zu publizieren.

Die Mitarbeiter erhalten von ihren Arbeiten zusammen 25 Freiexemplare.

Manuscripts should be addressed to/Envoyer les manuscrits à/Manuskripte sind zu senden an:

Prof. Dr. A. Brodal, Universitetet i Oslo, Anatomisk Institutt, Karl Johans Gate 47 (Domus Media), Oslo 1/Norwegen.

Prof. W. Hild, Department of Anatomy, The University of Texas Medical Branch, Galveston, Texas 77550 (USA).

Prof. Dr. R. Ortmann, Anatomisches Institut der Universität, 5 Köln-Lindenthal, Lindenburg.

Prof. Dr. T.H. Schiebler, Anatomisches Institut der Universität, Koellikerstraße 6, 87 Würzburg.

Prof. Dr. G. Töndury, Direktion der Anatomie, Gloriastraße 19, CH-8006 Zürich.

Prof. Dr. E. Wolff, Collège de France, Laboratoire d'Embryologie Expérimentale, 49 bis Avenue de la belle Gabrielle, Nogent-sur-Marne 94/France.

Ergebnisse der Anatomie und Entwicklungsgeschichte
Advances in Anatomy, Embryology and Cell Biology
Revues d'anatomie et de morphologie expérimentale

42 · 1

Editores

A. Brodal, Oslo · W. Hild, Galveston · R. Ortmann, Köln
T. H. Schiebler, Würzburg · G. Töndury, Zürich · E. Wolff, Paris

Henrik H. Lindeman

Studies on the Morphology
of the Sensory
Regions of the Vestibular Apparatus

With 45 Figures

Springer-Verlag Berlin Heidelberg GmbH 1969

Dr. H. H. Lindeman
E. N. T. Department · Ullevål Hospital
Oslo/Norway

ISBN 978-3-662-23030-5   ISBN 978-3-662-24992-5 (eBook)
DOI 10.1007/978-3-662-24992-5

© Springer-Verlag Berlin Heidelberg 1969. Library of Congress Catalog Card Number 65-20582
Titel-Nr. 6961.
Originally published by Springer-Verlag Berlin Heidelberg New York in 1969

Druck der Universitätsdruckerei H. Stürtz AG, Würzburg

Contents

I. Introduction

The membranous labyrinth lies enclosed in the very hard petrous part of the temporal bone and, as the name implies, its structure is extremely complex. This may explain why our knowledge of this organ and the vestibular sensory regions within it, is not yet satisfactory.

In recent years the rapid development of space research has provided a powerful stimulus to our interest in the vestibular apparatus. This has found expression in annual symposia, in which the role of the vestibular organs in the exploration of space is discussed. However, little is known as yet about the influence upon the equilibrial apparatus, during space flight, of weightlessness and other related conditions.

The inner ear has also acquired increased significance from an otosurgical point of view. Operations are today performed in regions previously inaccessible to surgery. This requires exact knowledge of anatomical details and of relations between the different structures in the inner ear.

However, the majority of illustrations available on the morphology of the vestibular apparatus are of little value for present day surgery or experimental investigations. The main reason for this is undoubtly the limitations of the histological methods used. Most studies on the vestibular sensory regions have been carried out on serial sections of decalcified and embedded temporal bones. Since usually a limited number of sections perpendicular to the epithelium are suitable for closer study, one can only study small areas of the vestibular sensory regions in each temporal bone. Furthermore, decalcification and embedding are time-consuming procedures and they are the cause of frequent artefacts, often misinterpreted as pathological changes. Orientation, a necessary prerequisite when a comparison of the same region in different animals is attempted, is likewise difficult in such sections, and the relation of certain cells to others is also lost.

Even though electron microscopical studies have shown that the structure of the vestibular sensory epithelium is considerably more complicated than was previously realized, such investigations can only be used to a limited extent in quantitative studies. Furthermore, only small areas are investigated by these methods, and the loss of orientation inevitable in studies under high-power magnification makes it difficult to carry out systematic investigation of definite areas. This is probably the reason why it has not been possible to follow up satisfactorily the study of regional differences in the structure of the sensory regions, described primarily by Lorente de Nó (1926) and Werner (1933, 1940). Differentiation of the sensory epithelium is seldom discussed, and it it still usual to regard the maculae as "flat spots" of uniform structure.

The present study is based on new methods of microdissection of the membranous labyrinth and the vestibular sensory epithelium. Using primarily these methods, an attempt has been made to: a) Demonstrate the gross anatomy of the membranous labyrinth and the peripheral branches of the vestibular nerve. b) Assess the form and size of the vestibular sensory regions. c) Investigate whether there is a special pattern in the vestibular sensory regions regarding the structure and innervation of the sensory epithelium and the structure of the statoconial membranes over the maculae. d) Carry out a quantitative estimation of the number and distribution of the two types of sensory cells.

For practical reasons, some of the investigations were made on only one species of animal. The temporal bone of the *guinea pig*, which forms the basis for all the studies, was used in these cases. To facilitate reading, a comprehensive account of material and methods is given separately (p. 8). In addition, a short review of the material and methods used in the special investigations is given in each chapter, which also includes a short discussion of the findings. A general discussion is found on p. 96[1].

II. Material and Methods

Approximately 200 guinea pigs, non albinos, weighing from 250 to 350 g were used in the investigation. All of them showed an obvious Preyer's reflex (twitching of the pinna to a sound stimulus).

These guinea pigs made up the bulk of the material. In addition the temporal bones from 6 rabbits, 3 cats, 4 squirrel monkeys, 4 human foetuses and 4 human adults were included in the study. The temporal bones from both sides were used.

A. Preparation, Fixation and Staining

In most cases the instruments and techniques described by Engström *et al.* (1966 b) were used to prepare the temporal bone for fixation and staining. Only the most important features will be mentioned.

The guinea pigs were decapitated without anaesthesia, the other experimental animals under sodium pentobarbital anaesthesia. The lower jaw was removed, the bullae tympanicae exposed and opened. After making a wide opening to the vestibule from the basal part of the cochlea, fixing fluid was injected into the vestibule, and the specimen was then placed in the fixative. In the guinea pig it took 1—2 minutes before both temporal bones were exposed to the fixing fluid. In the rabbit, cat and monkey, where the bone is stronger, the same procedure took 2—5 minutes.

The following fixatives and staining procedures were used:

1. Osmium Tetroxide (OsO_4). Most of the material was fixed and stained in a 0.5—1.5% solution of veronal-buffered osmium tetroxide for 1—3 hours in a refrigerator at about 4° C. After washing in physiological saline, the temporal bones were ready for further preparation. If they were not used the same day, they were transferred to 70% alcohol and stored in a refrigerator.

2. Methanol/Ether Fixation and Staining with Giemsa Solution. After fixing for at least 2 hours in equal quantities of methanol and ether, some temporal bones were stained for about 5 minutes in concentrated Giemsa solution. They were then transferred to xylol for further preparation.

---

1. Some preliminary results of the present study have been published previously by Engström *et al.* (1966a), Lindeman (1967).

3. Silver Nitrate. Without previous fixation the sensory regions in some animals were isolated in Ringer's solution and transferred to 0.5% $AgNO_3$ for about 5 minutes. After rinsing in distilled water, the specimens were exposed to ultraviolet light until they acquired a dark colour, by precipitation of silver chloride (see Romeis, 1948). Further preparation was carried out in xylol, after dehydration.

4. Formalin. A small number of temporal bones were fixed in 10% formalin and prepared without staining.

5. Formalin Fixation and Staining with Sudan Black B, according to the method described by Rasmussen (1961), was used in some cases.

6. Modified Maillet's (1963) Method. This was used in a great number of animals to study the innervation of the sensory epithelium. The fixing/staining solution consisted of a mixture of 1.5% osmium tetroxide solution and zinc iodide dissolved in distilled water (see Engström *et al.*, 1966 b).

All the temporal bones from humans, both adults and foetuses, were fixed by Dr. Bredberg in 1.5% osmium tetroxide solution (see Bredberg, 1968). The vestibular labyrinth was dissected by using the same technique as in animals.

B. Microdissection for Study of the Gross Anatomy of the Vestibular Labyrinth

These investigations were carried out on both decalcified and non-decalcified specimens. A number of temporal bones from guinea pigs and rabbits were decalcified in 5% HNO_3 until it was possible to remove the bone substance with forceps. In the guinea pig about 4 hours, and in the rabbit about 20 hours, were found to be satisfactory. After neutralizing in Na_2SO_4 the specimens were rinsed in water. Microdissection was carried out under the stereomicroscope either in distilled water or, after dehydration, in xylol. In xylol the bone structure is translucent, but the membranous labyrinth is also so transparent that it is easily damaged during preparation. By making the dissection in distilled water it was possible to isolate almost the whole of the membranous labyrinth (Fig. 1). In order to study small parts of it in detail, these were detached, mounted in Canada balsam or glycerin on a slide, covered with a cover glass and studied under an ordinary light microscope.

The general structure of the labyrinth was also studied in non-decalcified temporal bones from all species. Furthermore, a good view of the anatomy of the labyrinth was obtained when the vestibular sensory regions were dissected for detailed study.

C. Microdissection of the Vestibular Sensory Regions

Dissection was carried out on non-decalcified temporal bones under the stereomicroscope. Using powerful hooks and watchmaker's forceps, and sometimes a dental drill as well, the facial canal was opened and the facial nerve removed. The lateral wall of the vestibule together with parts of the walls of the osseous ampullae were then removed, thus exposing the saccule, the utricle and the three membranous ampullae (Fig. 2). This procedure took only a couple of minutes in the guinea pig, but slightly longer in the other species where the bone was harder.

As mentioned previously, a good idea of the anatomy of the membranous labyrinth was obtained during the dissection. More pronounced pathological conditions could also be seen with low power magnification (p. 55).

Further dissection was carried out under ×40 magnification with fine watchmaker's forceps. The saccule was opened first. Gentle squirts of fluid through a thin pipette were usually sufficient to separate the statoconial membrane completely from the sensory epithelium. Occasionally a pair of forceps was needed to help to free it. The statoconial membrane was then placed, under the fluid, on a cover glass which was lifted carefully out of the fluid and placed on a slide. It could now be studied more closely under a light and phase-contrast microscope, and X-rays could be taken (see p. 85).

The sensory epithelium was then separated from the subepithelial tissue. A very sharp knife was used for this purpose. The knife was introduced under the sensory epithelium outside its outer limits. It was easy to loosen large areas of the sensory epithelium in osmium-fixed preparations if one entered at the correct level. If the knife was introduced carefully under the sensory epithelium, it was possible to remove it almost completely in most cases. It was next transferred to a drop of glycerin on a slide, with the upper surface of the epithelium uppermost. This surface specimen was then covered carefully with a cover glass, ready for further study by light and phase-contrast microscopy. Epithelium that had not been fixed in osmium tetroxide was more difficult to separate from the underlying tissue and could therefore usually be studied only in fragments.

The macula utriculi was then prepared in the same way. In this case, however, the whole sensory region was released from its attachment anterior to the bone. It was not as easy to detach this sensory epithelium as that of the macula sacculi, because it was often difficult to fix the object during dissection. If, however, a good hold was obtained in the subepithelial tissue with forceps, it was nearly always possible to isolate the sensory epithelium almost completely and to transfer it to a slide for further study.

Finally the ampullae were dissected free, first the lateral then usually the anterior and finally the posterior. The roof and lateral walls of each ampulla were removed so that the crista ampullaris could be exposed. The cupula was removed and mounted on a slide for further study. The sensory epithelium was then separated in the same way as that of the maculae. By holding the specimen by the ampullar nerve, good fixation was obtained, and it was relatively easy to free the sensory epithelium in osmium-fixed preparations. The epithelium, which normally follows the saddle shape of the crista, is flattened once the cover glass is in the place.

Light and phase-contrast microscopy were carried out with a Wild M. 20 Research microscope. The sensory epithelium was viewed from above. Low power magnification ($\times 60$—100) was used to observe the form of the sensory epithelium, and certain structural peculiarities were also then apparent. Higher magnification ($\times 500$—1,000) was used for detailed study. By focussing at different levels it was possible to obtain optical sections through, for example, the hairs of the sensory cells, the surface of the epithelium and the nuclei of both sensory cells and supporting cells. Focussed at a given level, the structures could be seen clearly without troublesome interference from structures at other levels.

By moving the specimen, different regions entered the field of vision. In this way the whole structure of the sensory epithelium could be investigated rapidly, and the different regions related to one another.

Blood vessels and nerve fibres in the subepithelial tissue of the maculae were also studied directly after dissection. In the saccule this tissue, lying in a groove in the bone, was easily prepared. The subepithelial tissue of the macula utriculi was, as mentioned above, released before dissection of the sensory epithelium and could also be examined. So also could the subepithelial tissue be studied with the sensory epithelium intact over it. Although these specimens were often so thick and dark that detailed study was difficult, it was possible to correlate individual intraepithelial areas with corresponding subepithelial regions by focussing at different levels.

When it was desired to study other regions of the membranous labyrinth, these could be dissected free for direct study or embedded for section later.

D. Light and Electron Microscopy of Sections

Temporal bones from guinea pigs and squirrel monkeys were used for this purpose. They were fixed in cold 1.5% veronal-buffered osmium tetroxide, and stored for $1\frac{1}{2}$ hours in a refrigerator. After washing in Ringer's solution, they were dehydrated in increasing concentrations of alcohol. The vestibular sensory regions were dissected free in absolute alcohol, transferred to propylene oxide and embedded in Epon. Sections for light/phase-contrast and electron microscopy were carried out with a LKB Ultrotome. Sections for light/phase-contrast microscopy were stained with paraphenylene diamine, sections for electron microscopy with lead acetate, lead citrate or uranyl acetate. Electron microscopy was carried out with a Siemens Elmiskop 1 A.

Macroscopic pictures of the labyrinth were taken with a roll film camera, Hasselblad 500 C. equipped with a Zeiss S-Planar 120 mm lens, with extension tubes of varying length (2.1—150.0 cm). The microscopic pictures were taken under a Wild M. 20 Research microscope. Kodak Tri-X Pan 120 film was used.

E. Discussion

In the last century, anatomists relied to a large extent upon dissection for studying the structure of the inner ear. The most outstanding work of that period is Retzius' "Das Gehörorgan der Wirbelthiere" (1881a, 1884). Retzius combined the study of sections with microdissection of specimens fixed chiefly in osmic acid. Microdissection was of greatest value for investigating the gross anatomy of the labyrinth, but it was also used for detailed study of the cochlear and vestibular sensory regions and other parts of the membranous labyrinth. Retzius' plates thus show preparations of the vestibular sensory epithelium — where the pattern of sensory and supporting cells appears — from the macula sacculi of the alligator, from the macula utriculi of the pigeon and from the cristae of the rabbit and the cat.

Microdissection of the temporal bone was also used to some extent, for anatomical studies of the labyrinth, in the first half of the present century (for example Kolmer, 1927; de Burlet and Hoffmann, 1929; Werner, 1940), but not for direct study of the structure of the sensory epithelium. Similarly, in the last decade, microdissection under the stereomicroscope of parts of the membranous labyrinth, for embedding for electron microscopy, has given a good idea of the anatomy of the inner ear.

More recently the microdissection method has been resumed for investigation of the organ of Corti (Neubert, 1950, 1952, and others). In particular Engström and his group (Engström et al., 1962; Engström et al., 1966b; Kohonen, 1965; Bredberg et al., 1965; Bredberg, 1968), applying a special technique, have carried out systematic studies of the organ of Corti. Using osmium tetroxide and nerve-stained specimens, they have demonstrated the advantages of the method, both in experimental animals and in man, for the study of normal and pathological epithelium and for quantitative analysis of the cochlea.

This type of systematic study of surface preparations has apparently not been carried out in the vestibular sensory regions. Neumann and Neubert (1958), however, released the macula utriculi with its subepithelial tissue and studied the epithelium directly under the light microscope. However, probably because of the thickness of the specimen and the techniques used in fixation and staining, these studies provided no information about cell patterns, nor about the hairs of the sensory cells and other structural details. Engström et al. (1962) and Ades and Engström (1965) used microdissection for studies of both fixed and non-fixed vestibular sensory epithelium. However, Johnsson and Hawkins (1967), who used osmium tetroxide for fixing, considered that the maculae and cristae were too dark for surface studies.

The technique described in the present study for microdissection of the vestibular sensory epithelium should provide a useful supplement to the study of sections. The instruments required are few and inexpensive. With a little training, the investigator may find the vestibular sensory regions rapidly and easily, especially in the guinea pig and the human foetus, but the procedure is rather time-consuming in the adult human temporal bone. A good view of the anatomy of the labyrinth is obtained during dissection. Separation of the sensory epithelium from the subepithelial tissue needs training and some care, but it is surprisingly easy in osmium-fixed preparations. It is possible, for example, only a couple of hours after decapitating a guinea pig, to have the sensory epithelium isolated from all the cristae and maculae of one ear, for light and phase-contrast microscopy. The long waiting period and the artefacts which arise as a result of decalcification and embedding are avoided. Almost all the sensory and supporting cells in any sensory regions can be observed. These cells can be related to one another, and to other regions and other structures. Orientation in the specimen is certain, a factor of decisive importance when it is desired to compare corresponding regions in different animals. For these reasons, the method seems also to be especially suitable for quantitative assessment of both normal and pathological epithelia.

When desirable, other parts of the membranous labyrinth can be dissected in the same way as the cochlear structures (Engström et al., 1966 b), for direct study or for embedding and sectioning.

Osmium tetroxide, which was used in most of the studies in the present investigation, is the most common fixative used in electron microscopy of the structures of the inner ear. In addition to providing good fixation, it has several other advantages; the membranous labyrinth is seen very distinctly under the stereomicroscope; the sensory epithelium is easy to separate; and it is possible to study the hairs, the epithelial surface and the intraepithelial structures in the same specimen and under high magnification.

If it is not desired to study the specimens at once, they may be stored in 70% alcohol in a refrigerator. Storing for 1—3 days in alcohol does not seem to have any significant effect on the epithelial structure. Moreover, the sensory epithelium is easier to free, undamaged, from the underlying tissue. If, however, preparation is postponed for a longer time, the epithelium becomes so adherent to the subepithelial tissue that it may be difficult to separate it as an entity. It also becomes darker and the cell pattern is seen less distinctly in phase-contrast microscopy. Especially in the guinea pig where the crista has the shape of a fairly sharp ridge, the epithelium may be difficult to flatten, it often folds together and it is less easy to study. Sometimes it is difficult to see the cell pattern because of numerous granules in the epithelium. These are most obvious when the specimen has been stored for a long time in osmium tetroxide or in alcohol. It is often better to fix for shorter periods or in concentrations of osmium tetroxide solution lower than 1.5%.

Distortion of the sensory epithelium occurs occasionally. This is often due to the epithelium having been mounted in too little glycerin, and occurs most easily in specimens fixed for a short period. Distortion can, however, be recognized by a change in the cell pattern at the level of the epithelial surface, and by folding down of the hairs of the sensory epithelium. This type of surface specimen is not suitable for assessment of the area of the sensory region nor for estimation of cell density in different regions. The specimen can, however, be used for some structural studies and for estimation of, for example, the relationship between the numbers of the two types of sensory cells in the areas under consideration. In pathological epithelium where the degenerated sensory cells can be identified (Lindeman, 1967, 1969), it is also possible to estimate the relationship between the number of intact and of degenerated cells in the areas investigated.

The sensory epithelium, like other structures, can also be investigated without previous fixation/staining. Preparation of the sensory epithelium in these cases is, however, far more difficult, and only fragments of the sensory epithelium and detached cells are usually seen under the phase-contrast microscope.

III. Form and Interconnections of the Vestibular Ducts and Sacs

A. Introduction

The membranous labyrinth is a complicated system of interconnected, thin-walled ducts and sacs in which are located the sensory regions of the inner ear. It contains a fluid — the endolymph.

The first rudiments of the membranous labyrinth are ectodermal thickenings — the otic placodes — one on each side of the rhombencephalon. The otic vesicle (the otocyst) is formed by invagination of the otic placode. The otocyst gives rise to the inner layer of the walls in the membranous labyrinth. This layer is enclosed in a sheath of mesenchymal tissue. In-folding of the walls of the otocyst results in its division into three parts, which later develop into: 1. the endolymphatic duct and sac, 2. the utricle and the semicircular ducts and 3. the saccule, the ductus reuniens and the cochlear duct (see Bast and Anson, 1949).

The walls of the membranous labyrinth are thus composed of two layers: an inner epithelial layer of ectodermal origin and an outer mesenchymal layer. These are separated from each other by a basement membrane. The inner layer consists chiefly of a single layer of polygonal squamous cells. In some places, however, the epithelium is considerably more differentiated. This is especially true in the sensory regions — the organ of Corti, the macula sacculi, the macula utriculi and the cristae ampullares. In the areas surrounding these sensory regions the epithelium consists of cubic to cylindrical cells.

The membranous labyrinth lies enclosed in the bony labyrinth. The outer aspect of the fibrous layer of the membranous labyrinth is in contact with the periosteum of the bony labyrinth in some places, but is generally separated from it by a perilymphatic space. This space contains its own fluid — the perilymph. Thin connective tissue fibres or membranes which form a spider's-web-like network run together with blood vessels from the periosteum to the membranous labyrinth.

The membranous labyrinth was first described by Scarpa (1789). Names connected with relatively early studies on the structure of the labyrinth are Steifensand (1835); Schultze 1858); Odenius (1867)); Hasse (1867a, b), and, most important Retzius (1881a, 1884). The last author's comprehensive and systematic investigations, to a large extent based on microdissection, seem to be superior in quality to all subsequent studies on the general structure of the labyrinth.

More recently, other methods have predominated in the study of the anatomy of the inner ear. With the aid of so called corrosion methods, models of the bony labyrinth have been made. Gray (1909), using a special technique, was able to demonstrate the structure of both the bony and the membranous labyrinth, but his technique did not permit detailed study. Most studies on the anatomy of the labyrinth have been based on serial sections of decalcified and embedded temporal bones. Using this technique, however, it has often been difficult to obtain three-dimensional understanding of the structure of the membranous labyrinth; nevertheless it has provided valuable detailed information on the form and inter-relationship of the sensory regions (de Burlet and de Haas, 1923; Werner, 1933) and of the canals.

B. Material and Methods

Temporal bones from guinea pigs, rabbits, cats, monkeys and man (foetus and adult) were included in the investigation.

After fixation/staining, microdissection was carried out as described on p. 9. Several temporal bones from guinea pigs and rabbits were decalcified before this dissection. The isolated preparations of parts of the membranous labyrinth were mounted in glycerin on a slide or embedded in Epon for detailed study by light and phase-contrast microscopy.

Without decalcification it is only possible to prepare limited areas of the membranous labyrinth intact. After decalcification of the temporal bone almost the whole of the membranous labyrinth can be isolated (Fig. 1). This is a time-consuming job, necessitating great care. Both free preparations of parts of the membranous labyrinth and decalcified temporal bones were also embedded in Epon and sectioned for study by light and phase-contrast microscopy.

C. Observations and Comments

The present investigation showed that the structure of the membranous labyrinth was essentially the same in the different species examined. A few minor differences will be discussed.

When the stapes footplate is removed, access is gained to the vestibule, whose medial wall borders on the internal auditory meatus, its lateral wall bordering on the middle ear. In this relatively large space — also called the cisterna peri-lymphatica — the fine network of connective tissue threads and blood vessels that crosses the perilymphatic space in other parts of the vestibular labyrinth is lacking.

The saccule and the utricle are included in the vestibule (Figs. 2, 3a, b). *The saccule* is a flattened, irregularly shaped sac. It lies in a groove — the recessus sphericus — in the medial wall of the vestibule. Its upper part projects up to the utricle and has a broad attachment to the recessus utriculi, without there being any communication between the two cavities along the site of contact. Perlman (1940) and Igarashi (1964) described a clearly defined area where the wall of the saccule was thickened. They found this thickening of the wall in man and monkeys, but not in animals lower in the phylogenetic scale.

The sensory epithelium — macula sacculi — is covered by a statoconial membrane. The macula sacculi is not oval as described by Anson *et al.* (1967) and others, but rather like a hook, the anterior part of the epithelium bulging outwards in a superior direction (Figs. 2, 4, 19b). This bulging is also directed slightly

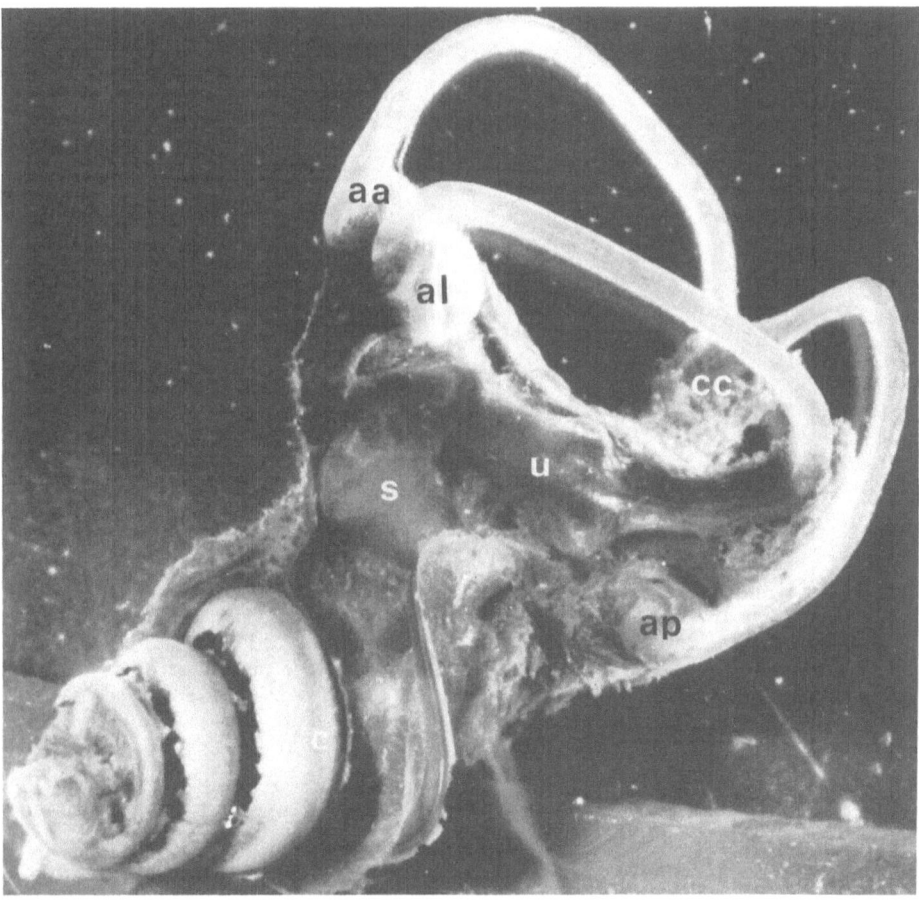

Fig. 1. Left inner ear of a guinea pig, seen from the lateral aspect and somewhat below. The membranous labyrinth is dissected free from the bony labyrinth after decalcification. The anterior and posterior semicircular ducts form the crus commune (*cc*), which opens into the utricle (*u*). The cochlea (*c*), forming $4^{1}/_{2}$ coils in the guinea pig, is seen on the left of the micrograph. *s* saccule, *aa* ampulla anterior, *ap* ampulla posterior, *al* ampulla lateralis (horizontalis). ×17

posteriorly. It has therefore been called the dorsal lobe (de Burlet and de Haas, 1924; de Burlet and Hoffmann, 1929; Werner, 1933). Since, however, the bulging is situated in the anterior region of the macula sacculi, and since it is clearly directed posteriorly only occasionally, the name dorsal lobe seems unfortunate and misleading.

The form and size of the macula sacculi varies to some extent from species to species, to a lesser extent within the same species. In the same individual the form of the sensory epithelium is practically identical on the two sides. In the normal anatomical position of the head, i.e. in man, the erect position, the macula sacculi is approximately vertical and the antero-superior part is so situated that the surface of the sensory epithelium also faces slightly inferiorly. Between the

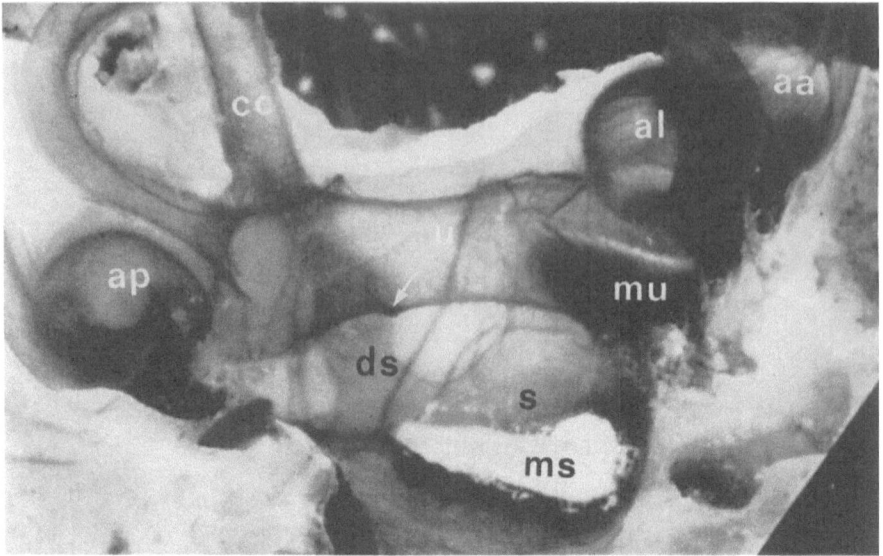

Fig. 2. The right vestibular apparatus in the guinea pig, as seen from the middle ear. The bone surrounding the vestibular sensory regions has been removed. In the saccule (*s*), the statoconial membrane of the macula sacculi (*ms*) is seen. The utricle (*u*) has a cylindrical shape. The macula utriculi (*mu*), seen slightly from below, is innervated by myelinated nerve fibres, which appear dark. In the ampullae, the saddle shape of the cristae is observed. *aa* ampulla anterior, *ap* ampulla posterior, *al* ampulla lateralis, *ds* saccular duct, *cc* crus commune, arrow: utricular duct. × 27

sensory epithelium and the bone there is a subepithelial layer of connective tissue, blood vessels and nerve fibres.

The saccule is connected with the cochlear duct by the *ductus reuniens* (Fig. 3a, b), a thin canal that leaves the postero-inferior part of the saccule. It runs along the bottom of the vestibule in a lateral direction and ends at the basal end of the *cochlear duct*. Hence this duct twists spirally for about $2^1/_2$ coils in man, and about $4^1/_2$ coils in the guinea pig (Fig. 1).

A canal leaves the posterior part of the saccule in a superior-posterior direction. It ends in the endolymphatic sac and receives a canal from the utricle. There have been considerable differences of opinion about the form and course of these canals, and the terminology has varied. In the present study Bast and Anson's (1949) terminology is used. According to this, the saccular duct extends from the saccule to the entrance of the canal from the utricle — the utricular duct. The endolymphatic duct starts where the two canals fuse (Figs. 5, 9) and ends in the endolymphatic sac. However, the saccular duct is considered by some to be a pouch of the saccule itself, and the utricular duct to be a direct connection between the utricle and saccule, thus justifying the name utriculo-saccular duct. Other authors are of the opinion that the endolymphatic duct leaves directly from the saccule proper and that the utricular duct ends in the endolymphatic duct, which therefore is called the utriculo-endolymphatic duct.

The saccular duct (Figs. 2, 5, 9) is a relatively short canal. It is very flattened at first, and the lumen then gradually increases to a sinus (sinus endolymphaticus).

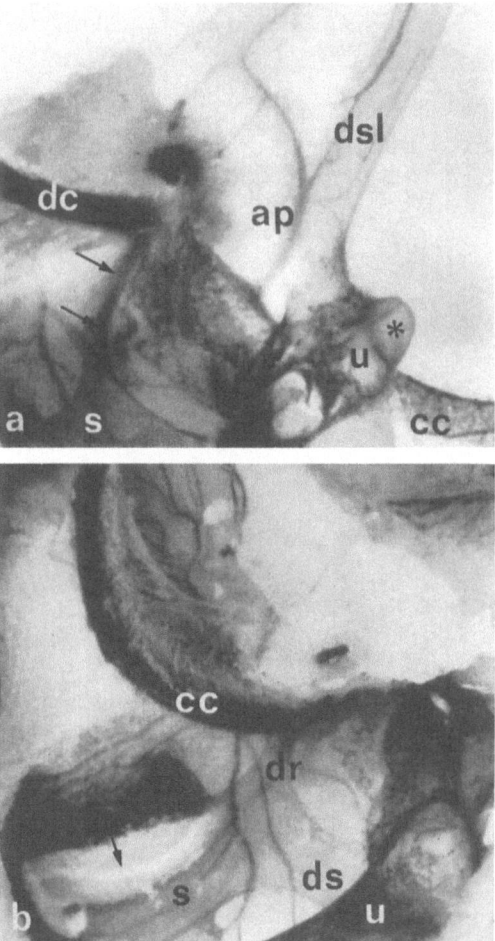

Fig. 3a and b. Photomicrographs of right inner ear of guinea pig. a Ductus reuniens (arrows) connects the saccule (*s*) and the cochlear duct (*dc*). * Points to a recess of the utricle (*u*), between the non-ampullary end of the lateral semicircular duct (*dsl*) and the crus commune (*cc*). *ap* ampulla posterior. b The relationship between the saccule (*s*), the saccular duct (*ds*) and ductus reuniens (*dr*) are shown. Note the snowdrift-like heapening (arrow) of the stato-conial membrane of the macula sacculi, corresponding to the striola. *dc* cochlear duct, *u* utricle. × 22

The maximal diameter of this sinus lies where the utricular duct enters it. Along that wall which is in contact with the periosteum, there are in the guinea pig obvious rugosities in the long axis of the canal.

The endolymphatic duct starts in the sinus endolymphaticus, at the meeting of the saccular duct and the utricular duct. It continues in the longitudinal axis of the saccular duct under the utricle, while the lumen gradually decreases. It soon leaves the vestibule and enters a bony canal, the aquaeductus vestibuli, which it fills completely. This part of the endolymphatic duct is called the isthmus and has a slightly curved course, postero-medially, to the crus commune. The

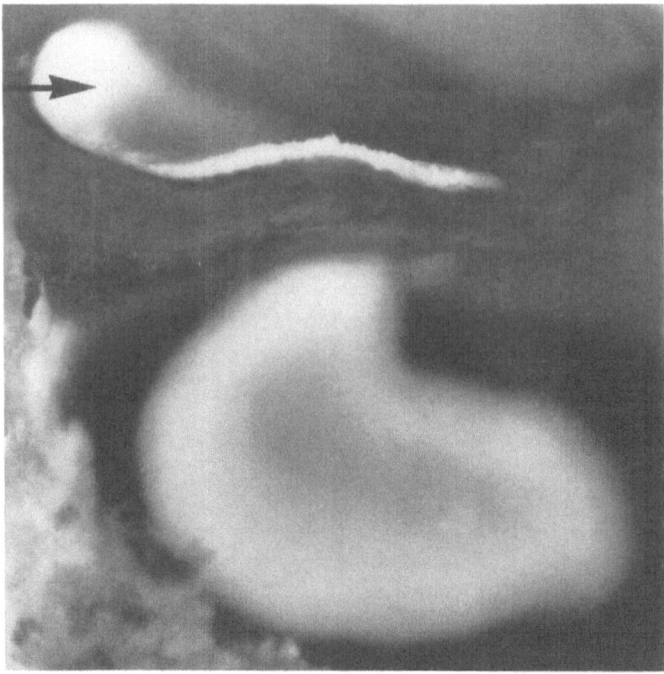

Fig. 4. Photomicrograph illustrating the relative positions of the statoconial membranes of the macula utriculi above and macula sacculi below in the squirrel monkey. In general, the surfaces of the two maculae are perpendicular to each other. The macula utriculi is lying more or less horizontal and the macula sacculi is standing vertically when the head is in its normal anatomical position in space. Note slight elevation of the anterior part (arrow) of the macula utriculi from the horizontal plane. × 68

canal ends at the posterior surface of the petrous part of the temporal bone in the endolymphatic sac, which lies between the two layers of dura mater. Guild (1927a) and Bast and Anson (1949) differentiated the endolymphatic sac into three regions based on morphological criteria. Its main functions are considered to be a resorptive action and a defence mechanism in the inner ear (Guild, 1927b; Andersen, 1948; Engström and Hjorth, 1950; Lundquist, 1965; Kimura and Schuknecht, 1965).

The *utricular duct* connects the utricle and the three semicircular ducts with the rest of the membranous labyrinth. In the literature this canal is nearly always shown as a canal of the same thickness as the saccular duct. The present observations show that it is much thinner (Figs. 2, 5, 9). It leaves the inferior aspect of the utricle, behind the recessus utriculi, through a narrow cleft-shaped opening in the bottom of the utricle. It usually remains close to the wall of the utricle and follows it in an arch. Before entering the sinus endolymphaticus, its lumen becomes more tube-shaped. There are, however, considerable variations in the course of the canal. In most cases it enters the sinus endolymphaticus at an obtuse angle to the endolymphatic duct, as found in man in 83.1% of cases by Bast and Anson (1949). However, it can also enter at a right angle or at an obtuse angle

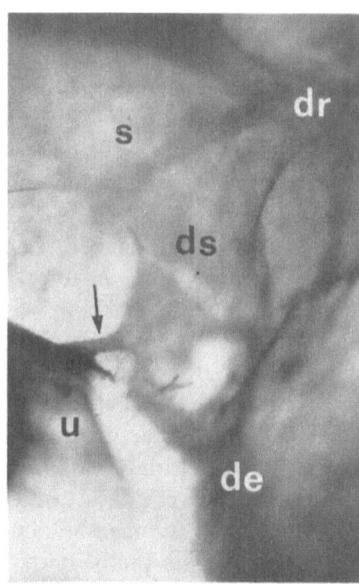

Fig. 5. Photomicrograph illustrating the relative size and position of the saccular duct (*ds*), the utricular duct (arrow) and the endolymphatic duct (*de*) in the guinea pig. The utricular duct is very thin, compared with the saccular and endolymphatic ducts. *s* saccule, *dr* ductus reuniens, *u* utricle. × 49

to the saccular duct. The form of the canal] also varies — it may be relatively short and thick, or longer and thinner. It is thicker in monkeys and man than in the guinea pig.

Not least from a functional point of view, the possibility of an utricular valve (utriculo-endolymphatic valve) has been discussed. The present observations support descriptions by Retzius (1884) and Werner (1940), of a thickened part of the utricular wall that has folded itself over the cleft-shaped opening of the utricular duct, thus forming a valve-like structure. Bast and Anson (1949) found that the two opposite walls of the cleft-shaped opening of the utricular duct are in contact with each other in most sections, and in other sections there is a space between them of only a few μ.

The utricle is an irregularly shaped tube with an oval cross-section; it is not shaped like a sac, as is often described. Fig. 2 shows the utricle in the guinea pig. In this animal it is very slender, but it is more irregular in monkeys and man. The utricle lies in the recessus ellipticus, an oblong groove in the medial wall of the vestibule, above the saccule. The sensory epithelium — the macula utriculi — is situated in the recessus utriculi, a dilatation of the anterior part of the utricle, which is not very prominent in the guinea pig, but is more distinct in man. The form of the macula utriculi is fairly uniform in the different mammals. Compared with the macula sacculi, it is relatively larger in man than in the guinea pig. It is almost kidney-shaped, the anterior part often being slightly broader than the posterior, with an in-pouching of the sensory epithelium medially (Fig. 19a). With the head in the erect position, the sensory epithelium in man lies with its main plane horizontally, but the anterior part is slightly elevated (cp, Fig. 4). It is covered by a statoconial membrane of the same shape. The sensory epithelium rests on a cushion of connective tissue, blood vessels and nerve fibres. This cushion is fixed to the bone anteriorly. In most guinea pigs the wall of the utricle is heavily pigmented, a pigmentation that is localized to very definite regions. The perimacular zone — the area immediately surrounding the sensory epithelium — in the recessus utriculi — is always free from pigment (Fig. 6). It is considered by several authors that the cells immediately surrounding the sensory epithelium, as well as other places in the utricular wall, have a secretory function (Saxén, 1951; Dohlman and Ormerod, 1960; Smith, 1956).

The utricle is connected with three semicircular passages — the semicircular ducts. As pointed out as early as 1835 by Steifensand, there are great discrepancies regarding the nomenclature of these canals, as both the bony and the membranous passages are often described as semicircular canals. Wolff *et al.* (1957) consider, from a functional point of view, that "canal" is more adequate than "duct" to denote the membranous semicircular passages. In accordance with Nomina anatomica, 1966, in the present study the membranous passage is termed "the semicircular duct", differentiating it from the bony passage — "the semicircular canal".

At one end of each semicircular duct there is a bladder-shaped enlargement — the ampulla. The non-ampullar part of the anterior and the posterior semicircular ducts, the so-called vertical semicircular ducts, run together to form a common passage, the *crus commune* (sinus superior utriculi) (Figs. 1, 7, 9), which ends in the utricle. The utricle thus has six openings, one to each of the three ampullae, one to the non-ampullar part of the lateral semicircular duct, one to the crus commune and one to the utricular duct. At the entrance of the non-ampullar part of the lateral semicircular duct, there is often in the guinea pig a pouching of the wall of the utricle in a lateral direction (Fig. 3a).

The three semicircular ducts lie in bony canals of the same shape — the semicircular canals. The position of the semicircular ducts is eccentric in the canals, with the convex side close to the periosteum. Otherwise there is a perilymphatic space of rather variable depth between the ducts and the canal. This space is crossed by a fine network of fibrous tissue and blood vessels. Although the course of the three semicircular ducts is slightly irregular, it can be said that each of them lies in a definite plane — and that they are in three planes at right angles to each other. This is shown in Figs. 1 and 7, which show the semicircular ducts and their ampullae in the guinea pig. A large number of authors have studied the position in space of the semicircular ducts in different animals and in man. In man with the head in the erect position the lateral (horizontal) semicircular duct has an angle of about 30° in relation to the horizontal plane.

The membranous ampullae fill the bony cavity more than do the semicircular ducts. The perilymphatic space between the roof of the ampulla and the periosteum is also crossed here by a fine network of fibrous tissue and blood vessels. The basal part of the ampulla lies in contact with the bone. Hence, a septum of connective tissue, blood vessels and nerve fibres projects into the lumen of the ampulla. This septum is shaped like a crest — the crista ampullaris — and is covered by sensory epithelium. Over the sensory epithelium is a gelatinous sub-

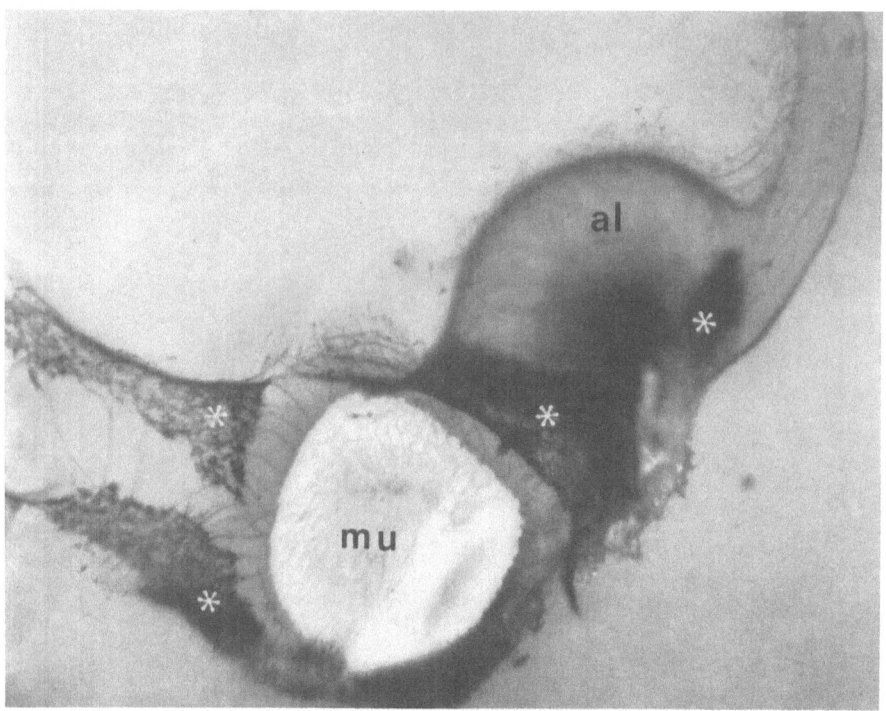

Fig. 6. The macula utriculi (*mu*) is covered by a statoconial membrane of the same size and form as the sensory epithelium. Pigment is localized in characteristic areas (*) in the walls of the utricle and the ampullae. Note perimacular zone without pigment. *al* ampulla lateralis.
× 58

stance — the cupula — which is normally considered to extend to the roof of the ampulla and out to its lateral walls (Steinhausen, 1933) (Fig. 8).

The crista ampullaris is reminiscent of a saddle, and extends across the ampulla. The crest is sharper in the middle regions than towards the side. It is blunter and relatively lower in man than in guinea pigs, where it reaches to about $^1/_3$ of the ampullar height. The present observations on whole preparations in guinea pigs showed that the sensory epithelium on the crista lateralis was rectangular with rounded corners (Fig. 12). The sensory epithelium on the cristae of the two vertical semicircular ducts had a more trapezoid form.

In the cat the sensory epithelium of the anterior and posterior cristae is divided in two by a transverse bar of cylindrical cells, referred to as the septum cruciatum (Gacek, 1961, and others). The term was originally introduced by Steifensand (1835), who observed a conical projection (processus laterales septi) from the cristae of each of the two vertical semicircular ducts in reptiles and birds. This gave the septum the shape of a cross — hence the name septum cruciatum (septum cruciforme). It is questionable whether the transverse bar on the cristae in the cat justifies this name.

The long axis of the crista is generally described as being vertical to the plane of the semicircular duct (Werner, 1940; Bast and Anson, 1949; and others).

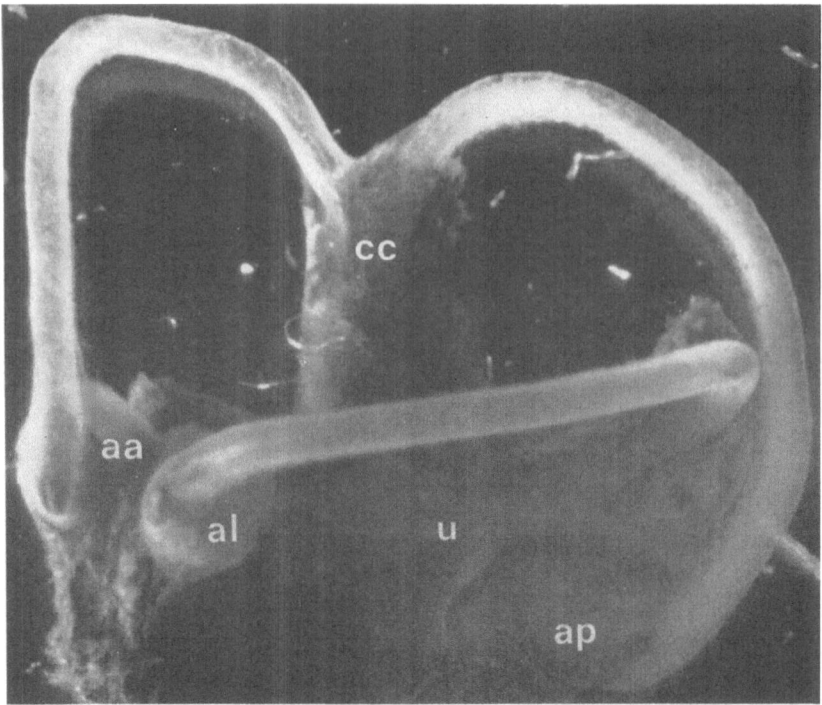

Fig. 7. The semicircular ducts are located in three different planes in space, perpendicular to each other. The non-ampullary ends of the anterior and posterior semicircular ducts form the crus commune (*cc*). *aa* ampulla anterior, *ap* ampulla posterior, *al* ampulla lateralis, *u* utricle. × 22

Present findings in the guinea pig have, however, shown certain irregularities in the position of the cristae, primarily in the anterior ampulla. In the first place it is not found at right angles to the plane of the semicircular duct; and it also lies in an asymmetric position basally, closer to one lateral wall than to the other. This asymmetry is found to a lesser extent in the position of the cristae in the other two ampullae. An asymmetrical position of the cristae has previously been described in birds (Steifensand, 1835) and fishes (Vilstrup, 1950), and is also seen in some of Retzius' (1884) illustrations in mammals.

Several authors have noted an asymmetrical distribution of the sensory epithelium on the two sides of the cristae. Wersäll (1956) found in the guinea pig that the sensory epithelium on the cristae of the vertical semicircular ducts extended further down on the utricular than on the canalicular side, whilst that on the lateral crista was symmetrically distributed.

The cells adjacent to the sensory epithelium are cubical-to-cylindrical and have a complicated structure (Kimura *et al.*, 1963). In the ampullar walls this epithelium is located in halfmoon-shaped areas called the plana semilunata (Fig. 8). The same name is often used incorrectly to describe the transitional epithelium which borders the sensory epithelium on the utricular and canalicular

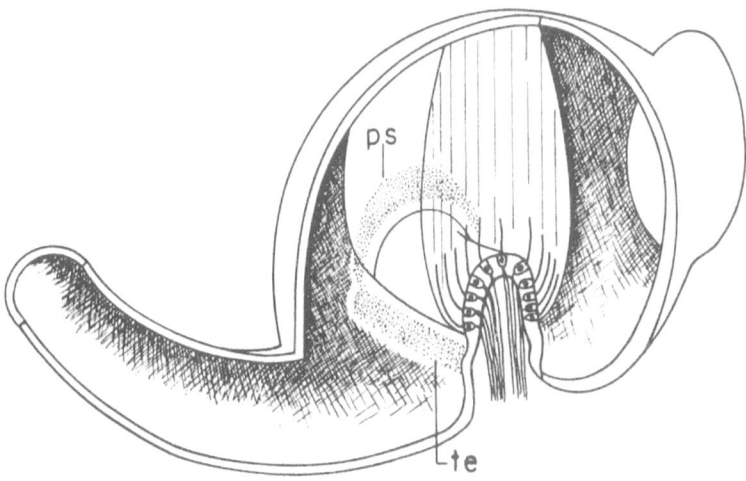

Fig. 8. Schematic drawing illustrating the architecture of the ampulla. The crista, traversing the ampulla, is covered by sensory epithelium. The hairs of the sensory cells protrude into the cupula, which is assumed to extend from the surface of the epithelium to the roof of the ampulla and outwards to the plana semilunata (*ps*) on the side walls of the ampulla. The transitional epithelium (*te*) is located at the base of the crista

sides of the crista (Fig. 8). Both the planum semilunatum and the transitional epithelium are considered to have a secretory function (Saxén, 1951; Dohlman and Ormerod, 1960; Bairati and Iurato, 1960; Kimura *et al.*, 1963; Dohlman, 1965; and others). The present study, however, does not permit any comment on these interesting observations.

The walls of the ampulla — semicircular duct system, like the walls of the utricle, contain pigment. In the ampullae this is situated chiefly in certain zones in the basal part (Fig. 6). The pigment cells are characterized by a number of interdigitating processes, obviously in intimate contact with capillaries. It has been pointed out that pigmentation is more marked in lower mammals than in man and that it has no connection with pigmentation of other parts of the body (Alexander, 1901 b). The pigment is probably melanin.

Lagally (1912), de Burlet (1920) and others have described a division of the perilymphatic space of the mammalian labyrinth into two parts, by a limiting membrane of connective tissue. The present observations in the guinea pig showed that this membrane was attached to the wall of the utricle and from here extended up to the uppermost part of the wall of the vestibule. By this connection the perilymphatic space surrounding the medial side of the utricle and the ampulla — semicircular duct system seems to be separated from the perilymph in the cisterna perilymphatica. As maintained also by Wersäll (1956) it is difficult to exclude the possibility that there is a communication between the two cavities. Kimura and Perlman (1956), in the guinea pig, after venous obstruction of the labyrinth, found haemorrhages in the pars superior of the labyrinth (around the semicircular ducts and utricle), but no bleeding in the rest of the perilymphatic space. This indicates that there is no communication between the two cavities.

IV. The Vestibular Nerve and Its Ramifications

A. Introduction

The sensory regions in the inner ear are innervated by the VIIIth cranial nerve which contains two functionally different components, the vestibular and cochlear branches, transmitting impulses from the organs of equilibration and hearing respectively.

The statoacoustic nerve leaves the brain stem laterally in the cerebello-pontine angle, below the inferior border of the pons. Close to it anteriorly are the facial and intermedius nerves. In man, these nerves run together into the internal acoustic meatus, surrounded by a common dural sheath. The facial nerve is located antero-superiorly; the intermedius and statoacoustic nerves lie below. The nerves separate at the bottom of the meatus. The facial and intermedius nerves then continue in their own canal. The cochlear nerve, which has its spiral ganglion in the modiolus, enters the organ of Corti. The vestibular ganglion (Scarpa's ganglion) is situated at the bottom of the internal acoustic meatus and is divided into a superior and an inferior part. The dendrites of the cells in the upper portion form the superior branch (the utriculo-ampullary nerve) which divides into the utricular and the anterior and lateral ampullary nerves. The inferior branch comes from the lower part of the ganglion and divides into the saccular and posterior ampullary nerves.

The above description of the peripheral branching of the VIIIth nerve was first modified by Retzius (1881a), who found that the macula sacculi in teleosts and reptiles also received fibres from the pars superior. Voit (1907) observed a similar double innervation of the macula sacculi in embryos of mammals, and his name has been connected with that branch of the pars superior which innervates the antero-superior part of the macula sacculi (Voit's nerve or anastomosis). These early observations were later confirmed by several investigators (Oort, 1918; de Burlet and de Haas, 1923; de Burlet, 1924; Alexander, 1924; Poljak, 1927a; and others).

Oort (1918) described, in addition, in mammals and man a connection between the pars inferior of the vestibular nerve and the cochlear nerve, the so-called vestibulo-cochlear anastomosis. There has been much disagreement about the type of fibres in this branch (see Lorente de Nó, 1926; Poljak, 1927a). Rasmussen (1946) showed by experimental studies that the anastomosis carried efferent fibres to the cochlea. These fibres could be followed to the superior olive and were both crossed and uncrossed. According to Rasmussen (1960) one-quarter of the fibres were from the homolateral side. However, some of the fibres in Oort's anastomosis did not degenerate after central severance of the efferent bundle, and Rasmussen (1953) assumed that these were afferent cochlear fibres.

Hardy (1934) observed in man, in serial sections, small bundles of fibres which came from the most basal parts of the spiral ganglion and went to the posterior part of the macula sacculi. She called this connection the cochleo-saccular nerve and assumed that the fibres represented a branch of the cochlear nerve. However, Rasmussen (1946), basing his findings on microdissection, concluded that these fibres consisted merely of an aberrant bundle from the main saccular nerve.

Shute (1951) made similar observations and he followed these fibres, in sections from human embryos, from the posterior part of the macula sacculi to the inferior vestibular ganglion (see also Weston, 1937).

Shute (1951) described also a small bundle of fibres from the lateral ampullary nerve to the utricular nerve. This bundle, which he called the superior utricular nerve, was assumed to contribute to the innervation of the lateral part of the macula utriculi

In addition many authors have described in various animals including cats, an area of sensory cells — the macula neglecta — at the bottom of the utricle, near the entrance of the crus commune. This region is innervated by a branch of the posterior ampullary nerve (see e.g. Gacek, 1961).

In addition to those mentioned above, several other fibre connections have been observed in the inner ear. An unknown number of intermedius nerve fibres, running in the vestibular nerve, leave this at the superior vestibular ganglion and cross to the geniculate ganglion (Alexander 1901a; Gacek and Rasmussen, 1961). A number of authors have described a connection between the two vestibular ganglia, the isthmus ganglionaris (see for example Alexander, 1899; Poljak, 1927a; Shute, 1951). Poljak (1927a) described in the mouse aberrant vestibular fibres in the petrous part of the temporal bone and Lorente de Nó (1926) observed a facio-cochlear anastomosis. Shute (1951) found that these fibres left the facial nerve at a level corresponding to the proximal part of the geniculate ganglion. It has been reported that the ganglion cells are not located exclusively in the vestibular ganglion but also lie scattered in the branches of the vestibular nerve and the isthmus ganglionaris (see for example Alexander, 1899; Poljak, 1927a; Weston, 1937; Wersäll, 1956). However, Ballantyne and Engström (1969) concluded that the vestibular ganglion cells in higher mammals form but one ganglion, and that scattered ganglion cells are seen only rarely at a distance from the main body of the ganglion.

B. Material and Methods

Temporal bones from guinea pigs, rabbits, monkeys and man were used for these studies. After fixation in 1.5% osmium tetroxide or in 10% formalin and staining with Sudan black B (see p. 9), the following methods were used.

1. Microdissection under the stereomicroscope and detailed study of isolated specimens under the light microscope.

a) After decalcifying the temporal bones of guinea pigs and rabbits in 5% HNO₃ (see p. 9), microdissection of the peripheral branches of the vestibular nerve was carried out in a bowl of distilled water. After dehydration, the isolated parts were transferred to xylol, mounted in Canada balsam and studied in detail under the light microscope.

b) Without previous decalcification the macula utriculi and macula sacculi with their subepithelial tissues were dissected free, and the distribution of the myelinated nerves was studied under the light microscope.

c) In order to study the innervation of the macula sacculi in more detail, temporal bones from guinea pigs and rabbits were placed for 1 hour in 5% HNO₃. Decalcification of the uppermost layer of bone in the recessus sphericus was thus achieved. This layer, together with the nerve fibres running therein, was preserved in one piece, together with the sensory epithelium and the subepithelial connective tissue. After dehydration and embedding in Canada balsam, details of the specimen were studied under the light microscope. With the aid of this technique any interference with the subepithelial course of the nerve fibres was avoided.

2. Embedding in Epon of decalcified temporal bones or pieces of tissue, and study of sections of various thicknesses under the light microscope.

C. Observations

1. General Course of the Peripheral Fibres of the Vestibular Nerve

Fig. 9 illustrates the peripheral branches of the vestibular nerve, and its connections with the cochlear and facial nerves.

The pars superior of the vestibular nerve (utriculo-ampullary nerve) gives off branches to the macula utriculi, lateral ampulla and anterior ampulla. A small bundle — Voit's nerve — close beside the utricular nerve, runs to the antero-superior part of the macula sacculi. More proximal fibres can be ollowed to the facial nerve; these fibres make up the facio-vestibular anastomosis mentioned above. In the rabbit a thin bundle of fibres can sometimes be demonstrated to run from the lateral ampullary nerve to the lateral part of the macula utriculi. A corresponding bundle is not present in the guinea pig. The pars inferior of the vestibular nerve divides into one branch to the macula sacculi and one branch to the posterior ampulla, and at this division, fibres also run across to the cochlear nerve — the vestibulo-cochlear anastomosis (Oort's anastomosis).

The myelinated nerve fibres innervating each crista ampullaris show a tendency to run in two branches — on the canalicular and utricular sides of the crista respectively. However, in the guinea pig these two branches are not completely separated.

2. Course of the Myelinated Nerve Fibres to the Macula utriculi

The macula utriculi is innervated by the utricular nerve from the pars superior of the vestibular nerve. The fibres pass anteriorly and slightly medially under the sensory epithelium and spread out underneath it. The fibres to the lateral areas show a tendency to run in an arch parallel to the other lateral border of the macula utriculi. This is especially evident in the guinea pig where, furthermore, the utricular nerve sometimes divides into two branches. One of these appears to innervate the striola and pars externa of the macula utriculi (see p. 43), the other the pars interna. However, there is a distinct overlapping of fibres from the two branches. In the rabbit and the monkey a small bundle leaves the utricular nerve and spreads out like a fan below a medially localized area of the macula utriculi.

3. Course of the Myelinated Nerve Fibres to the Macula sacculi

The macula sacculi is innervated by myelinated fibres from the pars superior and the pars inferior of the vestibular nerve (Figs. 9, 10). The thickest branch — the saccular nerve — from the pars inferior — passes under the sensory epithelium and supplies the main part of the macula sacculi. The antero-superior part of macula sacculi is innervated by a thinner branch — Voit's nerve — which comes from the pars superior. Studies of the borderline region between the areas of innervation of the two nerves in the guinea pig, rabbit, monkey and man show

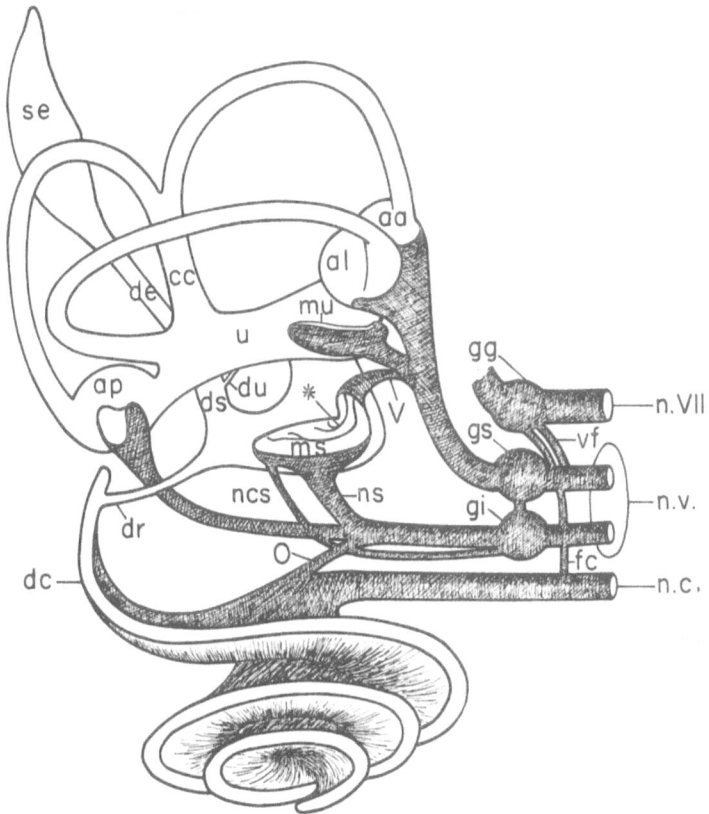

Fig. 9. Diagram showing the membranous labyrinth and the general plan of innervation of
the sensory regions in the inner ear in mammals. Note the innervation of the macula sacculi
(*ms*). In the guinea pig, bundles of thin myelinated fibres are observed (*) running from
Voit's nerve (*V*) into the main part of the macula sacculi. In man and monkey, the most
posterior tip of the macula sacculi is inconstantly innervated by a small bundle, termed the
cochleo-saccular nerve (*ncs*) by Hardy. In the diagram, the suggestion of Rasmussen and
Shute that this nerve is only an abberant bundle of the saccular nerve (*ns*) is followed.
mu macula utriculi, *aa* ampulla anterior, *ap* ampulla posterior, *al* ampulla lateralis, *cc* crus
commune, *ds* saccular duct, *du* utricular duct, *de* endolymphatic duct, *se* endolymphatic sac,
dr ductus reuniens, *dc* cochlear duct, *n VII* facial nerve, *nv* vestibular nerve, *nc* cochlear nerve,
gs superior vestibular ganglion, *gi* inferior vestibular ganglion, *gg* geniculate ganglion,
O anastomosis of Oort, *fc* facio-cochlear anastomosis, *vf* vestibulo-facial anastomosis

distinct overlapping of fibres from the two branches. This overlapping is most
pronounced in man.

Furthermore, in the guinea pig several small bundles of fibres are seen leaving
Voit's anastomosis and crossing over to the main part of the macula sacculi, where
they continue in its longitudinal axis (Fig. 10). These bundles consist of thin
myelinated fibres. They lie basally in the subepithelial tissue and only enter the
sensory epithelium when they terminate. Some of the fibres run under the striola,
but all seem to end superiorly to this in the pars interna of the macula sacculi
(see p. 46). In the guinea pig these fibres are seen quite regularly, but their

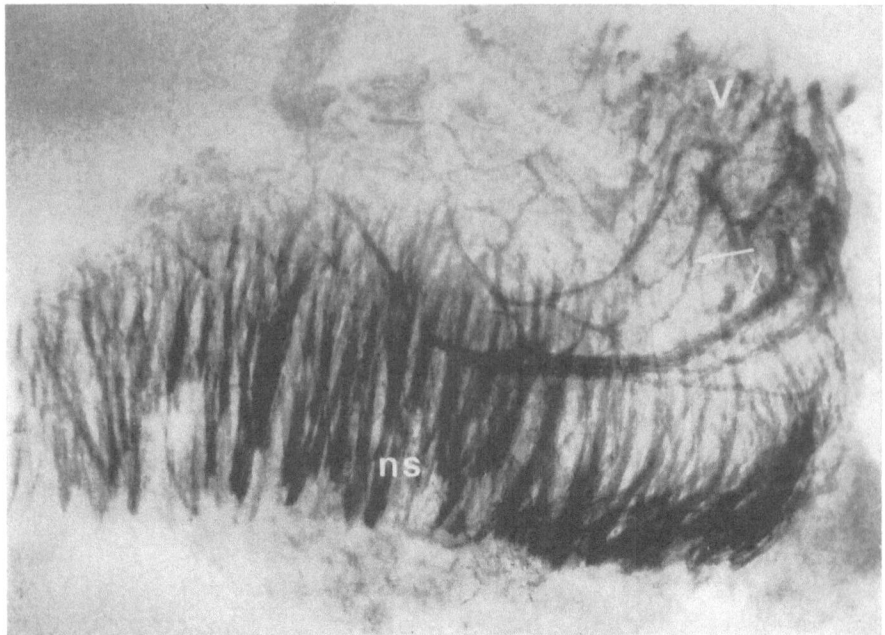

Fig. 10. Photomicrograph illustrating the innervation of the macula sacculi in the guinea pig. The macula sacculi is innervated by myelinated fibres from both the saccular nerve (*ns*) and Voit's nerve (*V*). Bundles of thin fibres (arrows) run from the latter into the main part of the macula sacculi. × 95

position and the number of bundles varies from one animal to another. In monkey and man it was not possible to determine whether similar fibres existed, since all specimens were too thick for detailed study. In the rabbit, however, a bundle of longitudinal fibres cannot be demonstrated under the main part of the macula sacculi. On the other hand, a large separate branch innervating the anterior part of the macula sacculi is seen here. This branch can be differentiated from the rest of the saccular nerve since it runs completely basally in the subepithelial tissue. Fibres from this bundle also seem to cross to the antero-superior part of macula sacculi, but it is difficult to follow the individual fibres over long distances in the dense subepithelial network of myelinated fibres. Proximally, as far as the inferior vestibular ganglion, it was not possible to differentiate these fibres from other fibres in the saccular nerve. Though less pronounced, the guinea pig, monkey and man showed a similar reinforcement of innervation of the anterior part of the macula sacculi.

In one monkey a small separate bundle was observed to innervate the most posterior part of macula sacculi. These fibres fused with the other fibres in the saccular nerve. It was not possible to observe a similar bundle in man, guinea pig, or rabbit.

D. Discussion

de Burlet (1924) emphasized that both Voit's connection and the saccular nerve innervated sharply defined areas of the macula sacculi in the guinea pig.

Poljak (1927a) also was of the opinion that the areas innervated by the two branches were relatively clearly demarcated even though he observed, in the mouse, some fibres from the saccular nerve which crossed over to the antero-superior part of the macula sacculi. On the other hand, he was not able to determine whether fibres from Voit's nerve crossed to the main part of the macula sacculi. The present observations, however, clearly show that there is a distinct overlapping in the border area, where fibres from Voit's and the saccular nerve cross the line of demarcation. This finding was most pronounced in man.

Furthermore, in the guinea pig, fibres from Voit's nerve cross to the main part of the macula sacculi, where they continue in its longitudinal axis. These fibres apparently have not been described previously. The reason for this may be that it is difficult to follow them in sections, while they are easily localized in surface preparations. However, these fibres do not seem to be constantly present in mammals. Thus they could not be demonstrated in the rabbit, and it was not possible to draw any definite conclusions from studies of monkey and man. On the other hand, the fibres were present regularly in the guinea pig, although both their position under the sensory epithelium and their number varied slightly from animal to animal. It is important to realize that the fibres are located completely basally in the subepithelial tissue. They are therefore easily damaged during preparation.

Using an acetylcholinesterase technique, Nomura et al. (1965) demonstrated fibres from Voit's nerve, in a position partly corresponding to that of the fibres mentioned above. The fibres are therefore possibly efferent. The present study has shown that they are thin, with a diameter of about one-third that of the thickest myelinated fibres. Referring to this it is interesting that Gacek (1960) has shown that the efferent vestibular fibres in the cat are relatively thin, most of them having a diameter of $2—3$ μ.

In the rabbit, a large branch was seen to take part in the innervation of the anterior part of the macula sacculi. This branch seems to correspond to Poljak's (1927a) bundle Y of fibres described in the mouse, rat and cat. In the guinea pig, monkey and man there is also often, though to a lesser extent, a similar reinforcement of innervation of the anterior part of the macula sacculi.

The small separate bundle which in monkey participated in the innervation of the posterior tip of the macula sacculi together with the saccular nerve seems to correspond to Hardy's cochleo-saccular nerve. The fibres, which occurred in 40% of the human temporal bones investigated by Hardy (1934), could be followed back to the inferior vestibular ganglion by Rasmussen (1946) in the dog and cat, and by Shute (1951) in man. They are therefore probably aberrant fibres from the saccular nerve, as assumed by Rasmussen (1946), and it should be stressed that the "cochlear ganglion makes no contribution whatsoever to the innervation of the saccule" [Shute, 1951 (loc. cit. p. 1017)].

Lorente de Nó (1926) and Poljak (1927b) observed that thick and medium-sized fibres in each ampullary nerve ran in two bundles, on the utricular and the canalicular sides of the crista respectively, which independently innervated the two halves of the sensory epithelium on the crista. On the other hand, both Wersäll (1956) and the present author have shown that there is no real separation in guinea pigs, since relatively thick fibres cross from one bundle to the other during their course in the crista.

Fibre counts in the vestibular nerve have given very variable results concerning the number of fibres. Thus, Gacek and Rasmussen (1961) found between 7,093 and 10,027 fibres in the guinea pig, whereas in man the number varied between 14,200 and 24,000 (Rasmussen, 1940). Wersäll (1956) also observed large variations in the number of fibres in the posterior ampullary nerve in the guinea pig, with values between 841 and 1,511 fibres. Determinations of the calibre spectrum have, in additon, shown that the fibres in the vestibular nerve generally have a larger diameter than those in the cochlear nerve. In man, Engström and Rexed (1940) found that 88.5% of the fibres in the vestibular nerve were between 2 and 9 µ thick, 7.2% were between 8 and 13 µ, while 4.2% of the fibres were less than 2 µ thick. In the guinea pig Gacek and Rasmussen (1961) found that, distal to the vestibular ganglion, 51% of the vestibular fibres had a diameter of 2.5 µ or less, 42% were between 2.5 and 5 µ and 7% between 5 and 8 µ thick. These values agree on the whole with Wersäll's (1956) findings in the posterior ampullary nerve.

The present investigations were confined to the peripheral branches of the vestibular nerve. The central projection of these fibres is, however, of great functional interest. Several authors have found that fibres from the vestibular sensory regions can be followed to very definite areas of the vestibular ganglion (Alexander, 1899; Poljak, 1927a; Shute, 1951; and others). Lorente de Nó (1926) divided the ganglion into 5 regions. Fibres from the anterior and lateral cristae, and from region b of the macula utriculi (an area which corresponds on the whole to the striola, see p. 53) could be followed to large ganglion cells in the superior ganglion (Lorente de Nó's pars magnocellularis anterior). Fibres from the rest of the macula utriculi went to small ganglion cells in the superior ganglion (pars parvicellularis anterior). The posterior ampulla was projected on to an area with large ganglion cells in the inferior ganglion (pars magnocellularis posterior a); fibres from the antero-superior part of the macula sacculi could be followed to another group of large ganglion cells in the inferior ganglion (pars magnocellularis posterior b); while the rest of the macula sacculi was projected on to an area with small ganglion cells in the inferior ganglion (pars parvicellularis posterior).

Weston (1939) found that the fibres from the cristae ampullares could be followed to larger ganglion cells than the fibres from the maculae. Otherwise there appears to be no more relevant information on the projection of the maculae and cristae to different groups of ganglion cells. Finally, it should be mentioned that Werner (1960) found an irregular distribution of large and small ganglion cells in the dog, rabbit and guinea pig.

Most authors agree that it is possible to follow Voit's nerve to the inferior vestibular ganglion (Lorente de Nó, 1926, 1931; Poljak, 1927a). The proximal part of the nerve is assumed to correspond to the connection between the two ganglia, the so-called isthmus ganglionaris (Shute, 1951). Stein and Carpenter (1967), however, were of the opinion that Voit's nerve came from the superior ganglion.

The axons from the bipolar ganglion cells can be followed to the four ipsilateral vestibular nuclei (the superior, medial, lateral and inferior nuclei), and to the cerebellum and a smaller group of cells — the interstitial nucleus of the vestibular nerve (see Brodal *et al.*, 1962, for a further subdivision of the vestibular complex). However, as shown by Walberg *et al.* (1958), only certain areas of the four classical vestibular nuclei are supplied by primary afferent vestibular fibres.

It is also of functional interest to know whether the different maculae and cristae project themselves on to separate areas of the vestibular nuclear complex. Lorente de Nó (1933) followed the axons from the cells in the different regions of the vestibular ganglion and found that the primary afferent fibres could be divided into five groups. These were distributed to the different nuclei according to a definite pattern, but there was no clear selective termination of fibres from the different sensory regions, and he therefore concluded that "the semicircular canals and the maculae have partly different and partly common central representation" (loc. cit., p. 33). It is interesting that the recent observations of Stein and Carpenter (1967) largely agree with Lorente de Nó's findings.

V. Structure of the Vestibular Sensory Epithelia

A. Introduction

The vestibular sensory epithelia are situated on the macula sacculi, the macula utriculi and the three cristae ampullares. Already the early authors (Scarpa, 1789; Steifensand, 1835) were able to follow the nerve fibres to these regions, but Schultze (1858) was the first to give a description of the structure of the sensory epithelium. He described in elasmobranchs three different types of cells, and also observed sensory hairs projecting from the surface of the epithelium. Odenius (1867) made similar findings in man. In the same year, however, Hasse showed in birds (Hasse, 1867a), and in dogs and cats (Hasse, 1967b) that the sensory epithelium on the cristae and the macula utriculi contained only two different cell types, which, according to his description, correspond to the now recognized sensory cells (hair cells) and supporting cells. The nerve fibres made contact with the sensory cells, and it was also from these cells that the hairs projected. Retzius (1871) found that each sensory hair was not homogenous, but consisted of a number of very fine threads of different lengths. By analogy with previous findings in the organ of Corti, Pritchard (1876) described an increased density of the uppermost structures of the sensory epithelium, and he called this the reticular membrane. After systematic studies (Retzius (1881a, 1884) found that the basic features of the structure of the vestibular sensory epithelia were uniform in all vertebrates. Van der Stricht (1908) and Held (1909) later showed that the sensory hairs were not all of the same type, as previously believed, but could be differentiated into true sensory hairs and a flagellum.

The electron microscope immediately provided new opportunities for morphological studies of the vestibular sensory epithelia. Observations by Wersäll (1954) and Wersäll *et al.* (1954) indicated that the structure of the vestibular sensory epithelia was more differentiated than previously supposed. Wersäll (1956) was able to show that in the guinea pig the sensory cells on the cristae ampullares were of two types, which he called type I and type II. Type I cells were bottle-shaped and surrounded by a nerve chalice that enclosed almost the whole cell. Type II cells were more cylindrical and were innervated by a number of nerve fibres that made contact with the base of the cell with bud-shaped nerve endings. Sensory hairs, one kinocilium and a number of stereocilia, projected from the free surface of each sensory cell. In the same year Smith (1956) showed that macula utriculi of the guinea pig had a similar structure.

Following these pioneer studies a large number of authors have used the electron microscope as a basis for detailed observations of the ultrastructure of the vestibular sensory epithelia (Engström and Wersäll, 1958a, b; Wersäll, 1960, 1961, 1967; Bairati, 1961; Engström, 1961, 1965; Engström *et al.*, 1962, 1965; Iurato, 1962; Iurato and Taidelli, 1964; Ades and Engström, 1965; Spoendlin, 1965, 1966a, b; Spoendlin *et al.*, 1964; Flock, 1964; Wersäll and Flock, 1965; Smith, 1967; and others).

After a study of serial sections in guinea pig and rabbit, Werner (1933) showed regional differences in the morphology of the maculae. He found that the gelatinous substance which surrounds the crystals had a special structure (see p. 94) in a central, curved area of the statoconial membrane. He called this area the striola. Opposite this area of the statoconial membrane he showed peculiarities in the structure of the sensory epithelium. The nuclei of the supporting cells were completely basal in position and clearly separated from the nuclei of the sensory cells, which formed a separate layer above the former. Corresponding with the striola of the macula utriculi there were occasionally so-called cystic cavities in the epithelium.

Wersäll (1956) found, furthermore, that type I cells were chiefly present at the top of the cristae, and that the number decreased peripherally. Type II cells on the other hand were chiefly present in the periphery of the epithelium. Later Engström and Wersäll (1958a) mentioned that a central semilunar area on the macula utriculi, which corresponded to the striola, appeared to have a large number of type I cells. They could not, however, find any definite regional differences in the distribution of type I and type II cells on the macula sacculi. Spoendlin (1956), making similar studies, found a concentration of type I cells at the top of the cristae and centrally on the maculae, in areas that appeared to correspond to Werner's striola.

B. Material and Methods

1. Material

Temporal bones from guinea pigs, rabbits, cats, squirrel monkeys and man (foetus and adult) were used for the study of the structure of the sensory epithelia. The guinea pig was used for quantitative studies.

2. Methods

a) Methods for Studying the Structure of the Sensory Epithelia

After opening the vestibule (see p. 8), the temporal bone was prepared according to the following methods: Fixation/staining in OsO_4, fixation in methanol/ether and staining in Giemsa solution, the $AgNO_3$ method (see p. 9).

The studies were carried out on surface specimens (see p. 9), or by light and electron microscopy, after embedding and sectioning (see p. 10).

b) Methods for Quantitative Studies

α) *Estimation of Number of Sensory Cells in the Vestibular Sensory Regions.* The temporal bones were fixed/stained in 1.5% OsO_4 for 3 hours. After rinsing in physiological saline, the specimens were transferred to 70% alcohol and stored in a refrigerator for 10 hours. The sensory epithelium was then isolated in distilled water, using the technique described on p. 9, placed on a slide in glycerin and carefully covered with a cover glass. If the sensory epithelium had been significantly damaged so that it was not complete, or if there were visible distortions of the epithelium (see p. 12), the specimen was discarded.

Counting of the sensory cells and estimation of the surface area of the epithelium were carried out as follows:

The contours of the sensory epithelium were projected on paper with a Wild projection apparatus and drawn with a pencil. The outlines of the sensory epithelium were then determined with a planimeter, and the surface area of the sensory epithelium was calculated. The area of the whole sensory epithelium, as well as that of the central zone, the striola, was determined on the maculae. The striola was easily seen even under low power magnification (Fig. 11 b, d). However, since its limits were not clearly defined, the area could not be calculated with complete accuracy. Using low power magnification, the striolar areas were determined in three maculae sacculi and three maculae utriculi where the striolae were seen especially distinctly. This was then controlled under high power magnification. The mean area occupied by the striola of the macula sacculi was about 13.0% of the area of the whole sensory epithelium, whilst the corresponding figure for the striola on the macula utriculi was about 8.0%. These figures formed the basis for all calculations in the maculae concerning the number of sensory cells in and outside the striola.

On the cristae ampullares there was a considerable variation in the density of the sensory cells (the number of sensory cells per unit surface area) in different regions. It was therefore found advisable to divide the surface specimen of the sensory epithelium into three zones — a peripheral, an intermediate and a central zone. On the whole, there was little variation in density in different regions of the peripheral zone in the same specimen. The place where the cell density started to diminish was at the boundary between this zone and the intermediate zone. In the central zone the density of the sensory cells was low and nearly constant. The boundary with the intermediate zone was taken to be where the density increased. Determination of the area of these three zones and the percentage proportion of the area of the whole sensory epithelium represented by each zone, was carried out in surface specimens of the sensory epithelium from three cristae laterales. The mean area of the peripheral zone was about 32.5%, the intermediate zone about 38.0% and the central zone about 29.5% of the whole area. Since the relationship between the areas of the different zones seemed also to be about the same in the cristae of the two vertical semicircular ducts, it was also used for determining the number of cells therein.

Using 1,000 × magnification, the sensory cells in the different regions were counted. The number of cells in a field of 0.001016 mm², which was marked on the eyepiece, was determined. The cells that were either within or touching the left and upper limits of the field were included, while the cells touching the right and lower limits of the field were not included. An average of about 950 cells distributed over about 60 fields were counted in each macula, and about 650 cells distributed over about 40 fields on each crista. An attempt was made to select the fields so that all the areas were equally represented.

The total number of cells in each area was calculated on the basis of these counts and these determinations of area (Tables 1 and 2).

The sensory cells were visualized by focussing on the hair bundles and on the free cell surfaces. The hair bundles were usually more distinct than the free surfaces of the sensory cells. However, since the hairs were sometimes torn off during preparation of the specimen, the number of cells in each field was always checked by focussing on the surface of the epithelium as well.

Calculation of the number of sensory cells was carried out on five maculae sacculi, on five maculae utriculi and on nine cristae ampullares, three from each crista.

β) *Methods Used for Quantitative Determination of Type I and Type II Cells.* Observations were carried out on surface specimens from both maculae and cristae. Attempts to find a method for selective staining of one of the two cell types were unsuccessful. Differentiation

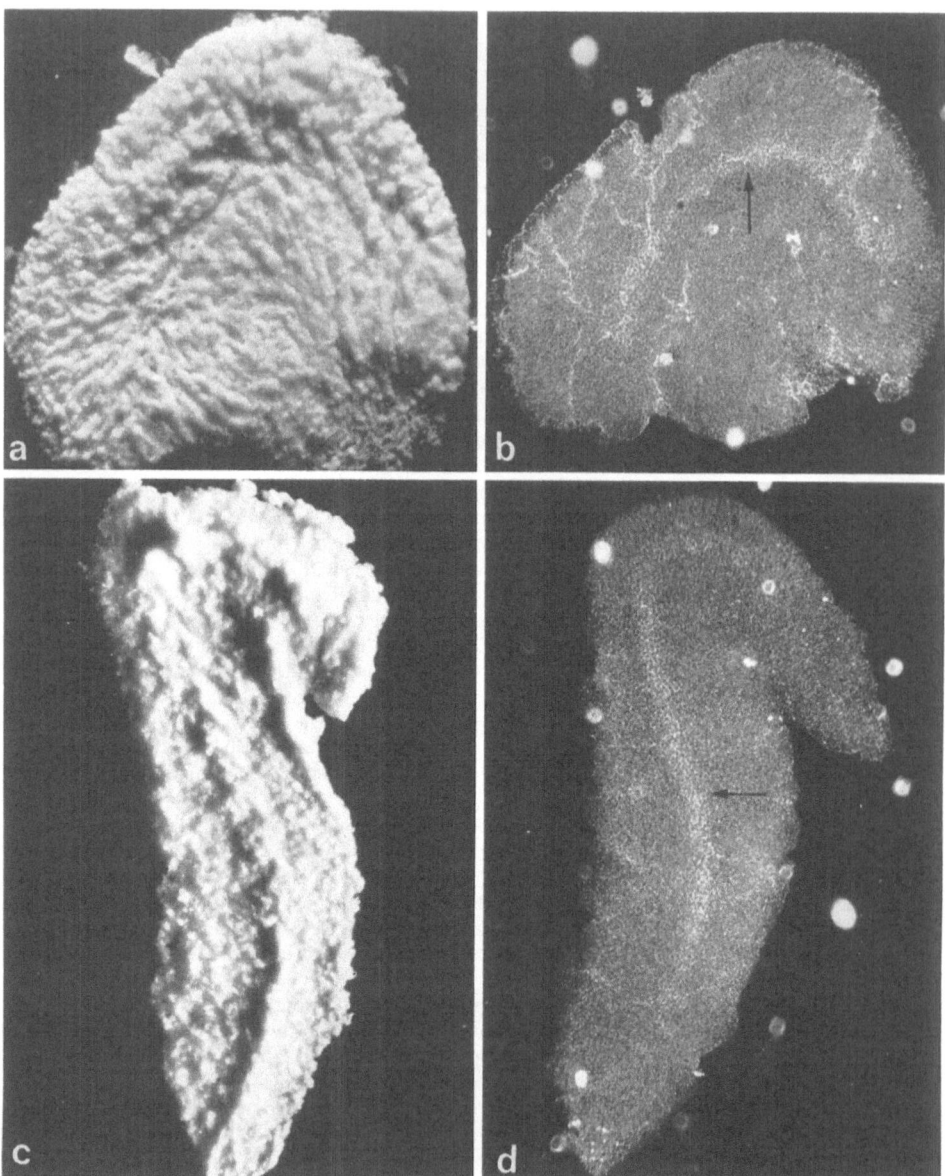

Fig. 11a—d. Surface preparations showing the statoconial membranes and the sensory epithelia of the macula utriculi (a, b) and the macula sacculi (c, d) in the guinea pig. The striola appears as a lighter zone running through both sensory epithelia (arrows). In the striola the statoconial membrane of the macula utriculi forms a groove, whereas in the macula sacculi it forms a ridge. a × 71, b × 60, c × 74, d × 76

of the sensory cells was therefore based on identification of the nerve chalices surrounding type I cells. In specimens fixed with osmium tetroxide, the mitochondria in the nerve chalices often make the chalice appear as a dark ring (Fig. 25). The nerve chalices, however, were not always equally distinct. The present investigations were therefore confined to specimens in which the nerve chalices could be seen distinctly. The nerve-stained specimens often showed good impregnation of the nerve chalices, especially in the peripheral regions of the cristae (Fig. 24a, b). It was in just these regions that it was difficult to differentiate type I and type II cells with osmium tetroxide fixation. Differentiation of the two cell types peripherally in the cristae was therefore based on studies of nerve-stained specimens. The number of type II cells in a field was found by subtracting the number of type I cells from the total number of sensory cells in the field.

To a lesser extent, differentiation of the two cell types was made with the electron microscope in sections taken parallel to the surface of the epithelium.

C. Observations

1. General Structure of the Sensory Epithelium

The basic features of the structure of the sensory epithelium are the same on the maculae and the cristae (Fig. 13). The epithelium consists of sensory cells and supporting cells, and it is separated from the subepithelial tissue by a basement membrane. There are numerous nerve fibres and nerve endings between the epithelial cells. The supporting cells extend from the basement membrane up to the epithelial surface. The sensory cells, on the other hand, extend from the surface of the epithelium down to a varying depth in the epithelium, but never as far as the basement membrane. Their nuclei are located higher than the nuclei of the supporting cells, which are generally found basally. Each sensory cell is provided with a bundle of sensory hairs, consisting of one kinocilium and a number of stereocilia.

There are two fundamentally different types of sensory cells, type I and type II. Type I cells have a bottle-like shape (Fig. 13). Beneath the free surface, the cell narrows to a neck, but it becomes wider basally, at the region of the cell nucleus. The cells are limited externally by a plasma membrane which continues around the sensory hairs. The stereocilia vary considerably in length, and are attached in a characteristic way to a cuticular plate at the free surface of the cell. In a cuticular free area there is a basal body from which originates a kinocilium, longer than the longest stereocilia. These sensory hairs will be considered in more detail in chapter VI.

The nucleus of the sensory cell is usually round or slightly irregular in shape. The cytoplasm has various types of organelles. Below the nucleus there is usually a characteristic accumulation of endoplasmic reticulum with attached and free ribosomes, and a few mitochondria. Above the nucleus a Golgi complex, a restricted number of mitochondria and usually some vacuole-containing bodies are seen. In the neck area of the cell, there is an abundant endoplasmic reticulum, and numerous small vesicles and filaments. Below the cuticular plate there is always a large collection of round or slightly oval mitochondria.

Except in the most apical region, type I cells are completely enclosed in a nerve chalice. This contains numerous mitochondria, neurofilaments and a few vesicles. One or more highly granulated, bud-shaped nerve endings are directly apposed to the outside of the nerve chalice.

Type II cells, which have a more cylindrical shape (Fig. 13), show no essential structural differences from type I cells. In the guinea pig their nuclei, especially on the maculae, are generally more superficially located than those of type I cells.

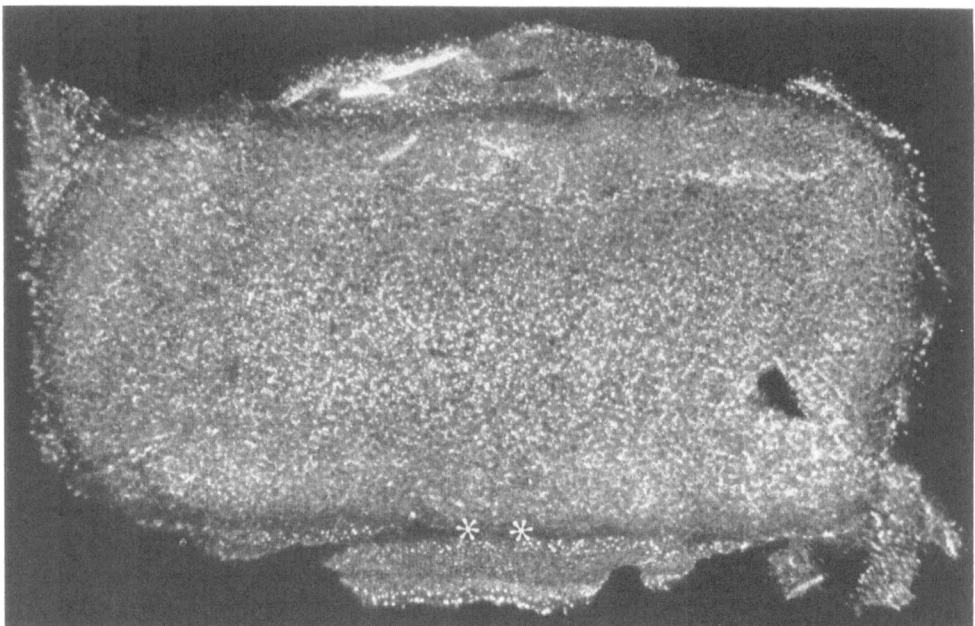

Fig. 12. Surface preparation showing the entire sensory epithelium of the crista lateralis in the guinea pig. * Indicates border between the sensory and the surrounding epithelium. The form of the sensory epithelium on the crista lateralis is almost rectangular with rounded corners. × 133

The infranuclear part of the type II cell may vary considerably in length and it often has an irregular shape. The basal parts of the cell are innervated by bud-shaped nerve endings of two types, richly- and sparsely-vesiculated.

Besides these two types of sensory cell, an intermediate type of sensory cell may also be observed, innervated by parts of a nerve chalice, and also receiving bud-shaped nerve endings directly.

The supporting cells enclose the nerve fibres in the basal regions of the epithelium; higher up they surround the sensory cells. On their free surface they have microvilli. In the apical region of the supporting cells a cytoplasmic condensation, forming the reticular membrane, is seen to be separated from the free surface by a thin layer of cytoplasm. The cytoplasm contains a Golgi apparatus and some mitochondria. In the more apical regions of the cells numerous slightly osmiophilic granules are observed. Further down in the cytoplasm there are, in the guinea pig, some much larger, strongly osmiophilic granules, up to 3 μ in diameter. The nuclei are often irregular in shape and usually contain several nucleoli.

The sensory and supporting cells form a characteristic pattern at the level of the surface of the epithelium. This pattern is seen particularly clearly in surface specimens (Figs. 14a, b, 15, 16, 21). The sensory cells have a round or oval free surface. They are surrounded by supporting cells, the free surfaces of which are more polygonal.

The number of supporting cells surrounding the sensory cells varies considerably (Figs. 14a, b, 16). There are generally less in the guinea pig than in cat or

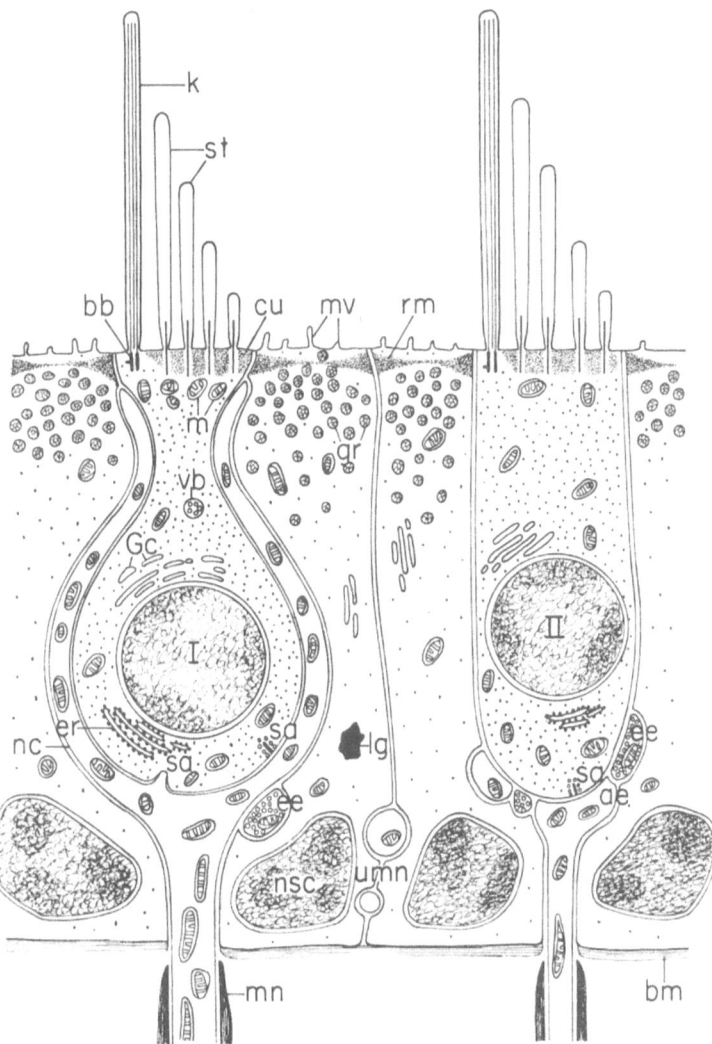

Fig. 13. Schematic drawing illustrating the general structure of the vestibular sensory epithelium. The flask- or amphora-shaped type I cell (*I*) is almost completely surrounded by a nerve chalice (*nc*). The more cylindrical type II cell (*II*) is innervated by bud-shaped nerve endings, which are of two types, much (*ee*) and sparsely (*ae*) vesiculated. Sensory hairs (*k* kinocilium, *st* stereocilia) protrude from the free surface of every sensory cell. *nsc* nucleus of supporting cell, *bb* basal body, *mv* microvilli, *cu* cuticle, *rm* reticular membrane, *gr* granules in the apical part of the supporting cells, *lg* strongly osmiophilic granule, probably lipid granule, *sa* synaptic area, *vb* vesiculated body, *m* mitochondria, *Gc* Golgi complex, *er* endoplasmic reticulum, *bm* basement membrane, *umn* unmyelinated nerve fibres, *mn* myelinated nerve fibre

man, in which usually 6 or 7 supporting cells are in contact with each sensory cell, less often 8, 9 or 10. Especially in the cat and in human foetuses regular constellations are seen, where the sensory cell in the centre is surrounded symmetrically by a rosette of supporting cells. Such a rosette figure is particularly common when the sensory cell is surrounded by a large number of supporting cells. As

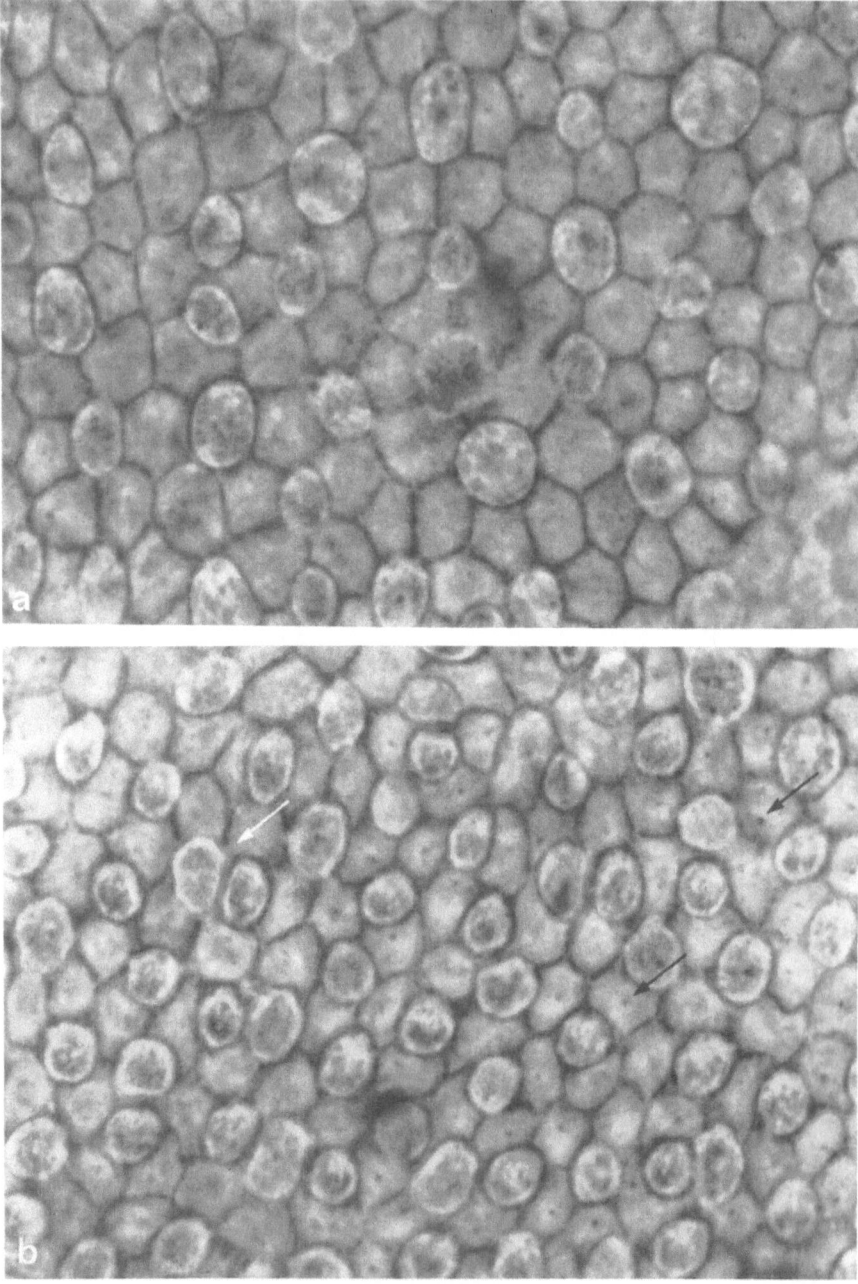

Fig. 14a and b. Surface preparation from the sensory epithelium of the crista lateralis in the cat. The microscope is focussed on the surface of the epithelium. The sensory cells have a round or oval free surface and are surrounded by a varying number of supporting cells. In the central parts of the crista (a), the density of sensory cells is less than in the periphery (b). Many of the sensory cells in the central areas have a large free surface, whereas in the periphery all the sensory cells bear a small free surface. Note the dark dots on the supporting cells (black arrows in b), corresponding to modified kinocilia or basal bodies. The white arrow indicates sensory cells lying close together. a and b × 1,560 shown in Figs. 14a, b and 16, however, the pattern of sensory and supporting cells in a given area is not completely regular, as for example, in the organ of Corti in the guinea pig and in human foetuses.

Not infrequently, two sensory cells are seen which lie close together at the level of the epithelial surface, and light and phase-contrast microscopy fail to show that they are separated by supporting cells (Fig. 14b, white arrow). Such cells are of different types, one type I and one type II cell.

Dark dots are seen on the free surface of the supporting cells (see e.g. Figs. 14b, 16). These dots probably represent basal bodies, possibly modified kinocilia. They are especially distinct in the cat and in the human foetus, and they seem to be present in these species on all supporting cells in the vestibular sensory epithelia. No special orientation was observed in regard to the position of these dark dots on the surface of the cells.

In normal vestibular sensory epithelia, so-called "collapse figures" are sometimes seen (arrow Fig. 17). Considerably increased numbers of these were seen in the guinea pig after application of ototoxic antibiotics (Lindeman, 1967, 1969). They must be regarded as degenerated sensory cells (see p. 54). In guinea pigs considered to be normal, up to four of these collapse figures could be found occasionally in one sensory region, but between nought and two such figures were commonly found. They were observed both in central and peripheral zones of the sensory epithelia. Although the structural features of the sensory epithelium are basically similar in all vestibular sensory regions, close investigation reveals distinct regional differences. These differences can be demonstrated not only between the maculae and the cristae, but also within each individual sensory region. Such structural differences are especially clearly seen when surface specimens of the sensory epithelium are studied.

2. Macula utriculi

Even under low-power magnification, a lighter zone running through the otherwise homogenous sensory epithelium can be seen in the guinea pig (Fig. 11b). The position of this zone corresponds essentially to Werner's (1933) striola. In the present study the *striola* is defined as the central, curved area where the sensory epithelium, and the statoconial membrane over this epithelium, have a characteristic structure. This area was seen in all the maculae investigated in mammals and man.

The main characteristic of the striola is the presence of sensory cells with a considerably larger free surface than the sensory cells outside the striola (c.p. Fig. 15). Thus several of the sensory cells at the surface of the epithelium in the striola of the guinea pig have a diameter of about 6 μ, while the mean diameter of the peripheral sensory cells is 3—4 μ. The striolar sensory cells with a large free surface are usually seen distinctly in surface specimens, and in the guinea pig they are mostly type I cells. In the striola there is also a smaller number of cells with a small free surface, which are not so distinctly seen in surface specimens. In the guinea pig these are usually type II cells, but in other species investigated this finding was not unequivocal. Outside the striola the free surface of the sensory cells is smaller and has a more uniform appearance. The type I cells in the striola, besides having a large free surface, are, in the guinea pig, very broad at the level

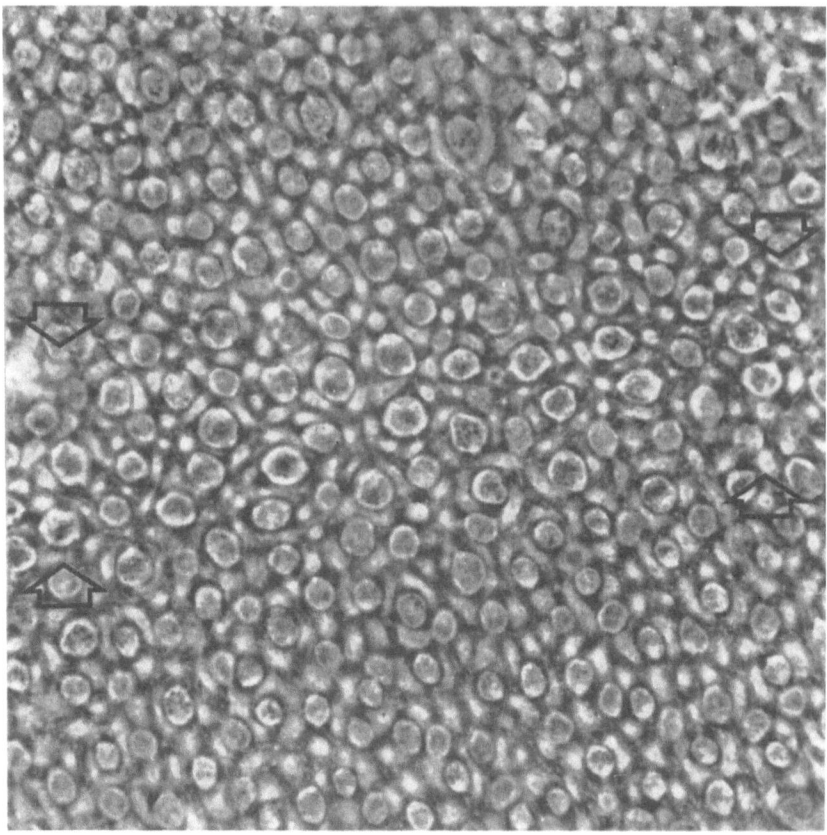

Fig. 15. Surface preparation from the macula sacculi in the guinea pig. In the striola, between the arrows, most of the sensory cells have a large free surface. These generally represent type I cells. The sensory cells outside the striola have a small free surface. The density of sensory cells in the striola is less than that on either side of it. × 846

of the nucleus. Particularly in the striola there are often several (2—5) type I cells in a common nerve chalice.

Obvious regional differences also appear when the microscope is focussed on the nuclei of the sensory and supporting cells. In the striola these nuclei are located almost in two separate planes, the nuclei of the supporting cells being completely basal, those of the sensory cells being higher in the epithelium (Fig. 18a, b). Peripherally in the maculae as well, the nuclei of the supporting cells generally have a more basal position than the nuclei of the sensory cells. In these areas, however, there is no obvious grouping of the nuclei of sensory and supporting cells at two distinct levels.

Even though the striola is seen in surface specimens, as a characteristic area, the transition between it and the peripheral regions is not distinct, and it is difficult to define its limits accurately. However, the striola is slightly broader in the posterior part of the macula utriculi than in the other regions, and the entire zone is narrower in the guinea pig than in the other species investigated.

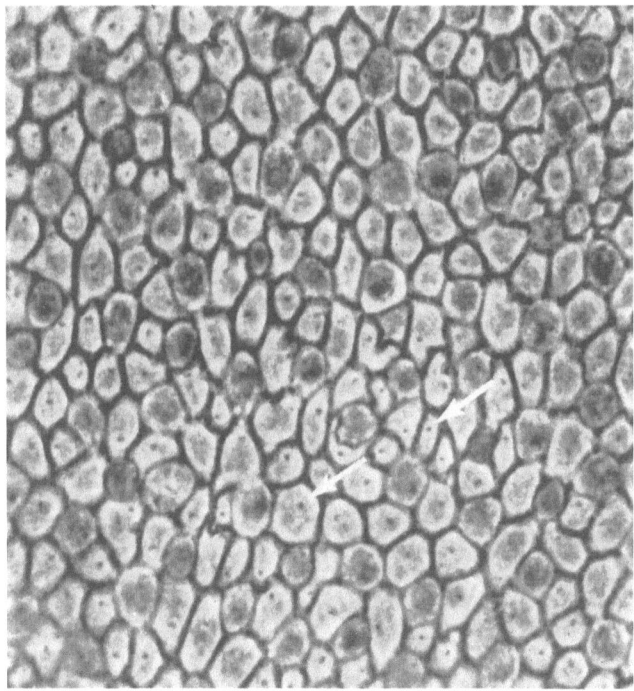

Fig. 16. Surface preparation from the macula utriculi of a human foetus. The sensory cells appear darker than the supporting cells. Note the dark dots on the supporting cells (arrows), representing modified kinocilia or basal bodies. ×1,095

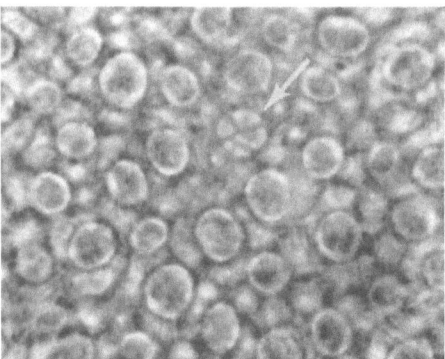

Fig. 17. Surface preparation from the macula sacculi of a "normal" guinea pig. A "collapse figure" (arrow), localized in the striola, is seen. This indicates a completely degenerated sensory cell. ×940

As shown in Figs. 11b, d and 19a, b, the striola does not extend right to the periphery of the sensory epithelium, but stops at some distance from its outer boundary. From here the cell pattern is similar to that found elsewhere in the periphery of the macula.

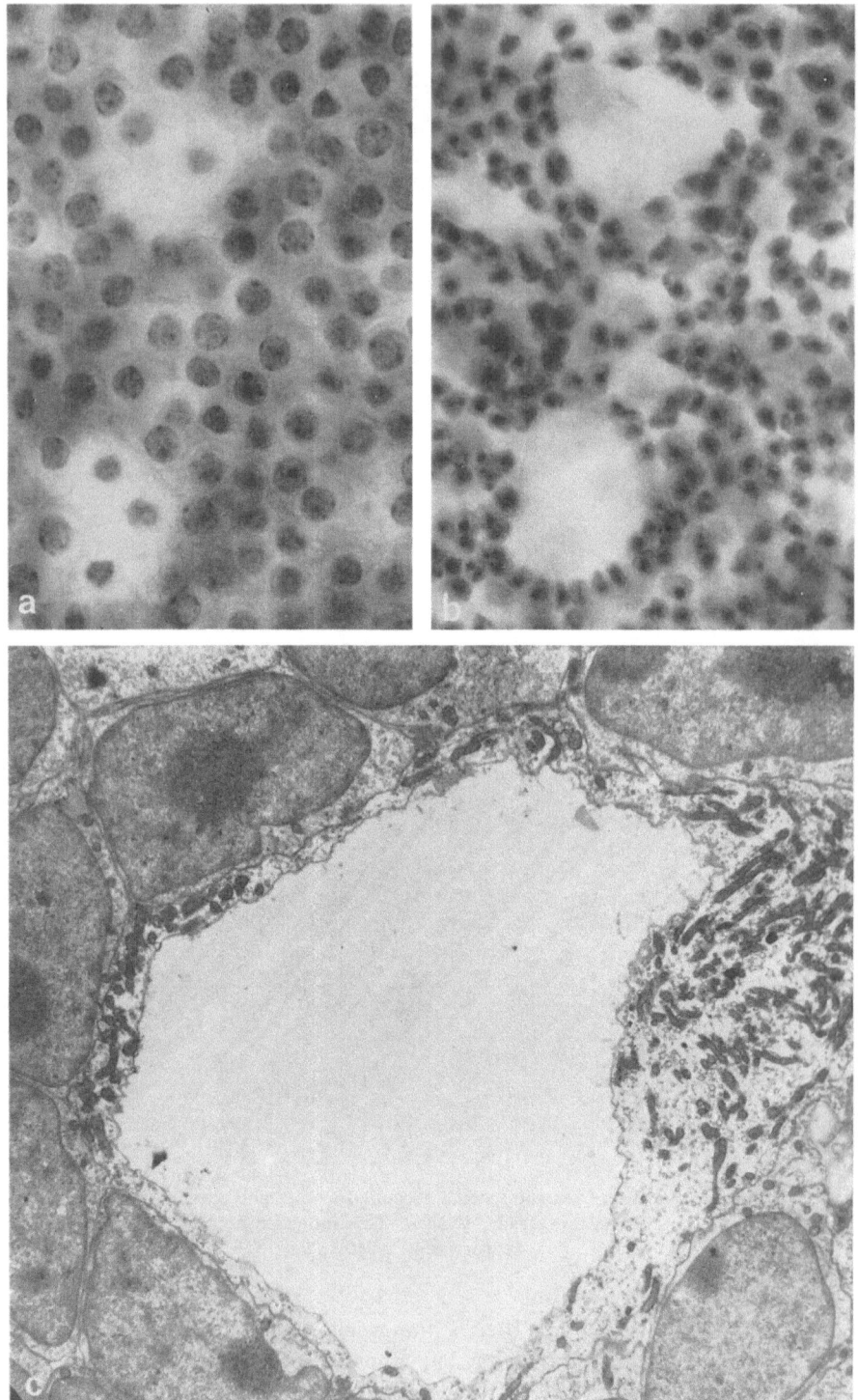

Fig. 18a—c

The vestibular sensory cells are morphologically polarized (see p. 59). Within relatively large areas of the macula utriculi, orientation of the sensory cells is fairly uniform. In the middle of the striola, however, there is a dividing line for the morphological polarization of the sensory cells, so that the cells on the two sides of this line have opposite orientation. The macula utriculi is thus divided into two regions — here called the *pars interna* and the *pars externa* (Fig. 19a, b). In the guinea pig these areas are of approximately the same size.

The medial border of the macula utriculi is invaginated in such a way as to give the macula a characteristic reniform (kidney-shaped) appearance (Fig. 19a). In this medial area the sensory cells lie scattered between the supporting cells (Fig. 20) (see also p. 61 and 84). The sensory cells in other regions are fairly sharply limited from the surrounding epithelium, but single sensory cells or groups of them are sometimes present outside the outer limits of the sensory epithelium.

The pattern of sensory and supporting cells often shows a characteristic orientation. The free surface of the supporting cells, and often that of the sensory cells, is oval. The long axis coincides with the direction of polarization in the area in question (Fig. 21). In certain regions, as in the lateral part of the macula utriculi in the guinea pig, this orientation is less distinct; and in the striola it is almost absent.

In the supporting cells of the guinea pig, in addition to numerous small apical granules, there are, as mentioned above (p. 36), a certain number of more basally-located, larger strongly osmiophilic granules. In the maculae these granules are up to 2μ in size. They show a characteristic accumulation in the striola (cp. Fig. 22b). The greatest concentration of granules is seen at the level of the nuclei of the sensory cells. In the sensory epithelium outside the striola, the number and size of these granules is much smaller.

Counts of supporting cells in guinea pigs and human foetuses show that there are more supporting cells than sensory cells. The number of supporting cells per unit surface area is slightly larger in the central regions (in, and just around, the striola) than in the periphery.

In the guinea pig, intraepithelial spaces are seen in the macula utriculi (Fig. 18a—c). These are almost exclusively confined to the striola, and they vary in size and shape. The largest space was over 38μ in diameter. These spaces are wider basally, narrowing towards the apex, and disappearing before the surface of the epithelium is reached. Fig. 18a and b show the appearance of the spaces in light micrographs. The nuclei of the sensory cells are located superficially (Fig. 18a), whereas those of the supporting cells lie basally (Fig. 18b), and are arranged in a rosette around the space. Fig. 18c is an electron micrograph of one of these spaces as it appears in a horizontal section. The cytoplasm lining the space contains vesicles, filaments and many mitochondria. It has not been possible to locate the nuclei of these cells.

Fig. 18a—c. a and b show a surface preparation from the striola of macula utriculi in the guinea pig. In a the microscope is focussed on the nuclei of the sensory cells and in b on the nuclei of the supporting cells. Note spaces in the epithelium. c Electron micrograph showing an intraepithelial space. The cytoplasm bordering the space contains numerous small mitochondria, vesicles and filaments. a and b × 780, c × 5,250

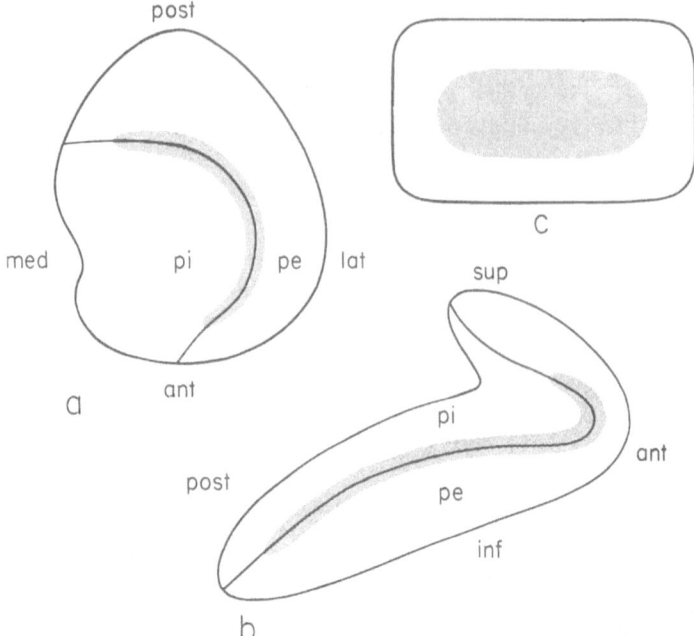

Fig. 19a—c. Diagram illustrating the subdivisions of the macula utriculi (a), macula sacculi (b) and the sensory epithelium of the crista ampullaris (c). Each of the maculae is divided into two areas, the pars interna (*pi*) and the pars externa (*pe*), generally with opposite morphological polarization of the sensory cells. The striola of the maculae and the central part of the epithelium of the crista are hatched

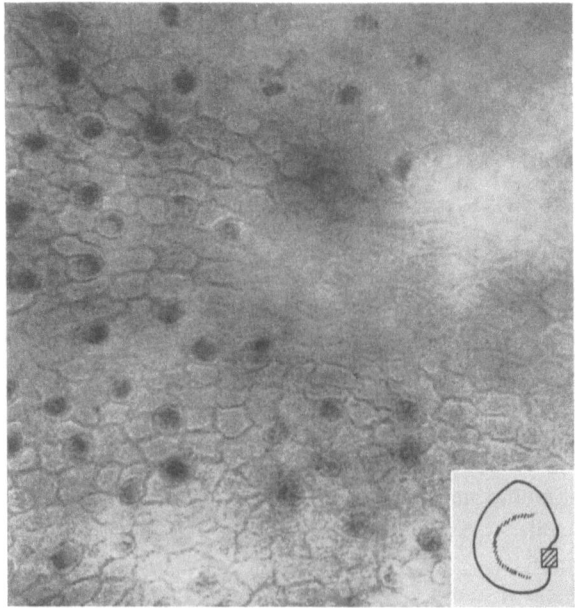

Fig. 20. Surface preparation from the macula utriculi of a human foetus. In a medially localized part of the macula utriculi (rectangle in schematic drawing), the sensory cells lie scattered among the supporting cells. × 780

The spaces are regularly present in the macula utriculi of the guinea pig. The number of spaces varies, however, in different animals. In one animal, phase contrast microscopy revealed 29 spaces, but usually the number was considerably smaller.

The area of 5 maculae utriculi in guinea pigs varied between 0,510 and 0.566 mm², with a mean value of 0.541 mm². The number of sensory cells was between 8,483 and 10,760, with a mean value of 9,260 cells (Table 1).

Table 1. *Number of sensory cells in the macula sacculi and the macula utriculi of the guinea pig*

Macula sacculi

striola = 13.0% of the whole area
periphery = 87.0% of the whole area

Specimen	Total area	Number of cells per surface area (0.001016 mm²)		Number of cells		Total
		striola	periphery	striola	periphery	
V 139 R	0.510 mm²	12.6	16.4	816	7,167	7,983
V 140 L	0.468 mm²	12.7	14.2	763	5,687	6,450
V 141 R	0.499 mm²	14.3	16.4	914	7,020	7,934
V 143 L	0.543 mm²	13.0	14.4	900	6,713	7,613
V 144 L	0.453 mm²	15.6	17.8	905	6,914	7,819
Mean	0.495 mm²	13.6	15.8	860	6,700	7,560

Macula utriculi

striola = 8.0% of the whole area
periphery = 92.0% of the whole area

Specimen	Total area	striola	periphery	striola	periphery	Total
V 69 R	0.510 mm²	14.9	18.5	596	8,560	9,156
V 140 R	0.526 mm²	12.8	17.2	528	8,192	8,720
V 141 R	0.561 mm²	12.7	17.3	562	8,618	9,180
V 143 R	0.540 mm²	12.8	16.2	546	7,937	8,483
V 146 R	0.566 mm²	15.3	19.7	678	10,082	10,760
Mean	0.541 mm²	13.7	17.8	582	8,677	9,260

The density of sensory cells (number of sensory cells per unit surface area) in the macula utriculi of the guinea pig shows regional differences. As indicated in Table 1, the density is smaller in the striola than at the periphery. There were only small variations within the entire striola and within the peripheral regions. The density of sensory cells was often greatest lateral to the striola, in the pars externa. In this region it was always slightly greater than in the posterior part of the pars externa. The mean density of sensory cells was approximately the same in the entire pars interna as in the pars externa. The mean number of sensory cells per field (0.001016 mm²) used for quantitative calculations varied slightly from one animal to another, from 12.7 to 15.3 in the striola, and from 16.2 to 19.7 in the peripheral regions.

The distribution of type I and type II cells is not the same in all regions of the macula utriculi. The striola contains an especially large number of type I cells.

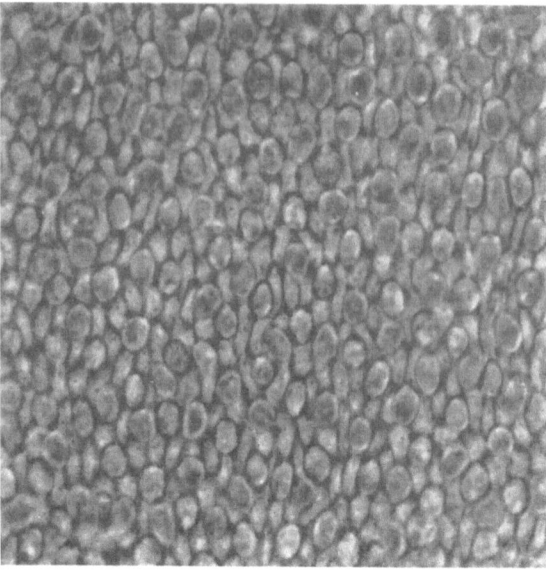

Fig. 21. Surface preparation of the macula utriculi in the cat. The supporting cells and to a lesser extent also the sensory cells, have an oblong free surface, the long axis of which corresponds to the morphological polarization of the sensory cells in the actual area. × 780

As shown in Table 3, there were on average about twice as many type I cells as type II cells in this region. Although the density of sensory cells in the striola is less than that outside this region, there are also more type I cells per unit surface area in the striola than there are peripherally. Outside the striola there were about equal quantities of each of the two cell types. The marginal zone of the sensory epithelium, however, seems to have a greater number of type II cells. Otherwise, no regional differences in the peripheral distribution of the two types of sensory cells were observed.

3. Macula sacculi

Low power magnification of the macula sacculi in the guinea pig shows that here also a lighter zone runs through the sensory epithelium (Fig. 11d). As in the macula utriculi this region, the striola, is characterized by especially large sensory cells (Fig. 15); in the guinea pig these are predominantly type I cells with a large free surface and a conspicuous increase in size at the level of the nuclei. Between these cells there is a smaller number of sensory cells of type II with a small free surface. Like the macula utriculi, the macula sacculi can be divided into two areas with a different polarization of the sensory cells, the pars interna and the pars externa (Fig. 19b). The dividing line for polarization lies in the middle of the striola (see p. 62). The pars externa is slightly larger than the pars interna. The striola occupies a larger percentage (about 13%) of the sensory epithelium on the macula sacculi than on the macula utriculi (about 8%). The width of the striola is constant except for the most anterior part where it turns off in a superior

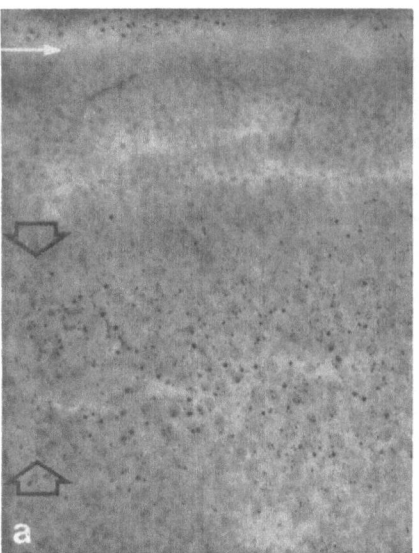 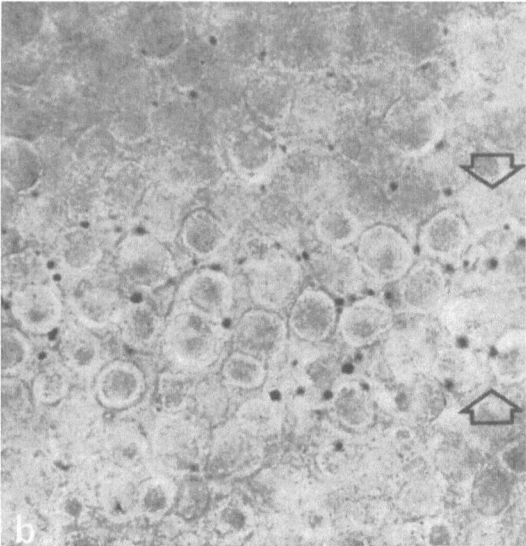

Fig. 22a and b. Surface preparations showing an accumulation of strongly osmiophilic granules (between arrows) in the central region of the crista lateralis (a) and in the striola of the macula sacculi (b). Similar granules are also observed in the epithelium surrounding the sensory regions. The white arrow indicates the border of the sensory epithelium of the crista. a × 186, b × 780

direction, parallel to the outer limits of the epithelium (Fig. 19b). Here the striola is slightly broader than elsewhere in the macula sacculi.

There are no significant differences between the macula sacculi and the macula utriculi with regard to the structure of the sensory epithelium, the cell pattern, and the distribution of nuclei of sensory and supporting cells in the striola and peripheral regions. The free surfaces of the supporting cells, and often also of the sensory cells, are oval with the long axis in the direction of polarization. This corresponds to the findings in the macula utriculi (see p. 43). The orientation is less conspicuous in the anterior part of the pars externa, and this pattern is lacking in the striola of the guinea pig.

The structure of the sensory epithelium in the antero-superior bulge of the macula sacculi, which is innervated by Voit's nerve (see p. 26), shows no difference from the epithelium elsewhere. In some guinea pigs, however, the sensory cells were less densely packed in this region than in the rest of the periphery of the macula sacculi, but this was not a constant finding. Within a small area, in the angle between the antero-superior bulge of the sensory epithelium and the rest of the macula sacculi (Fig. 19b), the density of sensory cells is likewise often small. In the macula sacculi there are also large, strongly osmiophilic granules in the supporting cells (Fig. 22b), the accumulation of these granules to the striola being even more characteristic than in the utricle.

The intraepithelial spaces were also seen in the striola of the macula sacculi, but they were fewer than in the macula utriculi.

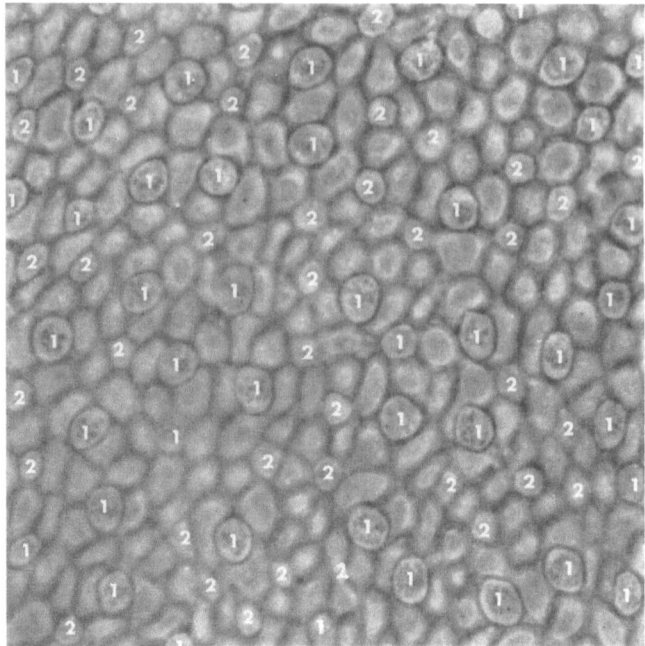

Fig. 23. Surface preparation from the central region of the crista anterior in the guinea pig showing the relative distribution of type I (*I*) and type II (*II*) cells. The sensory cells of type I have in general a large free surface whereas the type II cells have a smaller free surface.
× 780

The density of the sensory cells is lower in the striola than at the periphery and about the same in the pars externa and the pars interna. As shown in Table 1, the density of sensory cells in the macula sacculi appears to be lower than that in the macula utriculi. The mean number of cells per unit surface area varied in different animals, from 12.6 to 15.6 in the striola, and from 14.2 to 17.8 in the peripheral regions.

The area of the sensory epithelium was found to be between 0.453 and 0.543 mm², with a mean value of 0.495 mm². The number of sensory cells in the macula sacculi was between 6,450 and 7,983, with a mean figure of 7,560 cells.

As shown in Table 3, the distribution of type I and type II cells in the striola and in the peripheral regions was approximately the same as in the macula utriculi.

4. Sensory Epithelia on the Cristae ampullares

The structure of the sensory epithelia on the cristae ampullares shows obvious regional differences. In the central regions there is considerable variation in the size of the free surface of the sensory cells (Fig. 14a). Most of the sensory cells have a large free surface; in the guinea pig these are generally type I cells (Fig. 23). They are seen more distinctly in surface specimens than other cells with a small free surface, which usually are type II cells. In the cat this pattern is not obvious, many type II cells in this animal having a large free surface. In the peripheral regions of the sensory epithelium the surface area of the sensory cells is small and gives an impression of uniformity (Fig. 14b). The sensory cells in the centre of the cristae are shorter and broader than the sensory cells in the peripheral regions where they are more slender and often reach down almost to the basement membrane. No differences in cell patterns can be demonstrated between the utricular and canalicular sides of the same crista, nor can differences be shown when corresponding regions of different cristae are examined.

While the pattern of sensory and supporting cells on the maculae generally shows obvious orientation, which coincides with the direction of polarization in the area, this cannot be shown with certainty on the cristae by focussing on the epithelial surface. The number of supporting cells on the cristae is greater than the number of sensory cells and is slightly higher in the central regions than peripherally.

The supporting cells of the cristae in the guinea pig also have large strongly osmiophilic granules. In the central regions the concentration and size of these granules even exceed that in the striola of the maculae (Fig. 22a). The largest granules observed were more than 3 μ in diameter. The position of the granules varies. Some are seen at the level of the nuclei of the sensory cells, but they have generally a more basal position. In the peripheral regions a few granules of similar type are observed, but they are much smaller than those seen centrally.

Table 2. *Number of sensory cells on the cristae ampullares of the guinea pig*

peripheral area   = 32.5% of the whole area
intermediate area = 38.0% of the whole area
central area   = 29.5% of the whole area

Specimen	Total area	Number of sensory cells per surface area (0.001016 mm²)			Total number of sensory cells
		periphery	inter-mediate area	central area	
Crista anterior					
V 145 L	0.355 mm²	21.4	16.6	8.9	5,545
V 151 L	0.393 mm²	20.3	14.8	8.0	5,644
V 152 R	0.360 mm²	20.4	14.3	8.3	5,136
Mean	0.369 mm²	20.7	15.2	8.4	5,442
Crista posterior					
V 146 L	0,382 mm²	20.3	16.7	9.6	5,916
V 148 L	0.336 mm²	22.8	18.3	10.3	5,760
V 150 L	0.292 mm²	22.1	16.1	9.8	4,615
Mean	0.337 mm²	21.7	17.0	9.9	5,430
Crista lateralis					
V 145 L	0.384 mm²	20.4	14.7	9.0	5,622
V 146 L	0.386 mm²	20.3	14.9	9.2	5,690
V 151 L	0.380 mm²	20.8	16.0	8.5	5,753
Mean	0.383 mm²	20.5	15.2	8.9	5,688

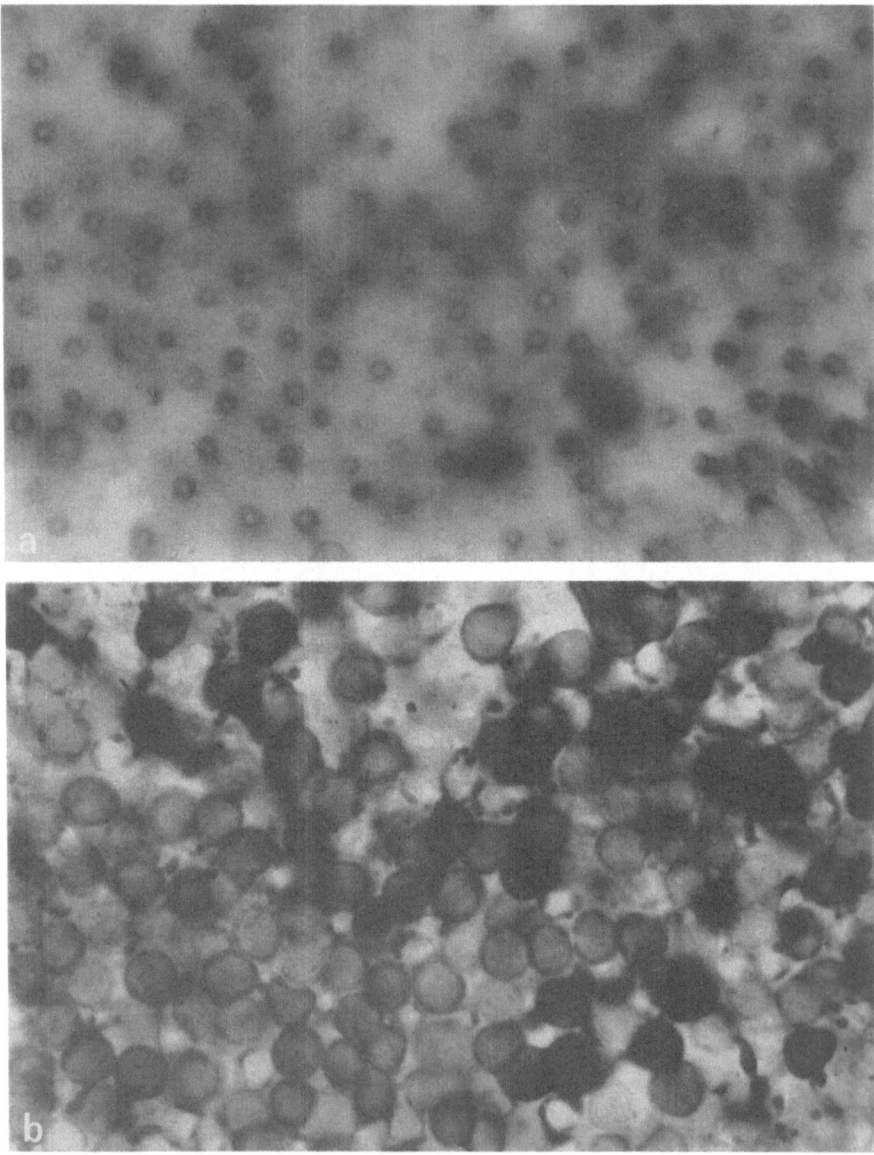

Fig. 24a and b. Nerve stained surface preparation from the periphery of the crista epithelium in the guinea pig. The type I cells are densely packed. In a the microscope is focussed on the neck of the nerve chalices. In b the microscope is focussed on the body of the type I cells in the same region. × 780

The regional differences in the density of sensory cells on the cristae are very striking (see Table 2 which shows the findings on the different cristae). Centrally, in an area occupying a little less than one-third of the total region of the sensory epithelium, the density is almost constant in the same crista and lower than that in the striola of the maculae. In a peripheral zone, occupying about one-third

Table 3. *Number of type I and type II cells in the vestibular sensory regions of the guinea pig*

Striola				Periphery			
Specimen	Type I	Type II	Number of cells counted	Specimen	Type I	Type II	Number of cells counted
Macula sacculi							
V 69 R	63.4%	36.6%	153	V 69 R	47.9%	52.1%	140
V 97 L	70.8%	29.2%	72	V 138 L	52.1%	47.9%	73
V 142 R	66.0%	34.0%	209	V 142 R	48.4%	51.6%	188
Mean	66.7%	33.3%	434	Mean	49.5%	50.5%	401
Macula utriculi							
V 69 R	68.2%	31.8%	135	V 69 R	52.2%	47.8%	299
V 144 R	65.8%	34.2%	76	V 138 L	47.7%	52.3%	398
V 153 L	66.4%	33.6%	128	V 144 R	51.1%	48.9%	131
Mean	66.8%	33.2%	339	Mean	50.3%	49.7%	828
Crista anterior							
V 147 R	62.6%	37.4%	107	V 135 R	57.1%	42.9%	240
V 148 L	55.7%	44.3%	106	V 136 L	61.2	38.8%	204
V 149 R	62.0%	38.0%	100				
Mean	60.1%	39.9%	313	Mean	59.1%	40.8%	444
Crista posterior							
V 140 R	60.3%	39.7%	141	V 137 R	58.1%	41.9%	43
V 146 L	63.0%	37.0%	92	V 139 L	65.7%	34.3%	35
V 150 L	58.2%	41.8%	98				
Mean	60.5%	39.5%	331	Mean	61.9%	38.1%	78
Crista lateralis							
V 144 R	65.8%	34.2%	111	V 81 R	63.0%	37.0%	92
V 147 R	65.1%	34.9%	169				
V 149 R	62.5%	37.5%	120				
Mean	64.5%	35.5%	400	Mean	63.0%	37.0%	92

of the epithelial area, the density is also almost constant, with the exception that the sensory epithelium adjacent to the planum semilunatum shows slightly lower density than the epithelium peripherally on the canalicular and utricular sides. The greatest density of sensory cells in the vestibular sensory regions was found in this area. The transition between the peripheral areas and the central regions was occupied by an intermediate zone, covering a little more than one-third of the area of the sensory epithelium. The cell density gradually decreased in a central direction. Table 2 shows the area of the sensory epithelium on the different cristae, and the number of sensory cells. This varied between 4,615 and 5,916. The mean figures were 5,442 sensory cells on the crista anterior, 5,430 on the crista posterior and 5,688 on the crista lateralis.

The relationship between the number of type I and type II cells in the different regions is shown in Table 3. Both in the central regions of the sensory epithelium

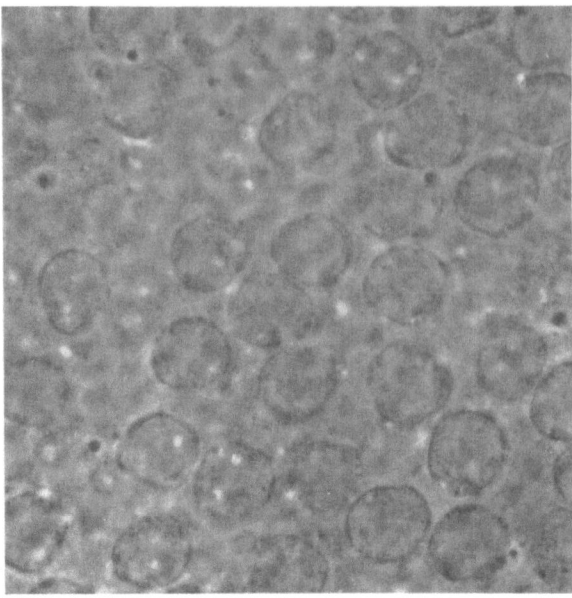

Fig. 25. Surface preparation from the central region of the crista lateralis in the guinea pig. Nerve chalices appear as dark rings and allow identification of type I cells. Osmic acid fixation. × 780

and peripherally (which includes the peripheral and intermediate zones) there are more type I than type II cells. No definite pattern in the distribution of the two types of sensory cell can be seen, and the relationship between type I and type II cells is approximately the same over the entire crista. There are, however, far more type I cells peripherally per unit surface area (Fig. 24a, b), since the cell density is much greater here than in the central region. A narrow marginal zone with predominantly type II cells seems, however, to be more pronounced on the cristae than on the maculae.

Several type I cells with a common nerve chalice can also be seen on the cristae, but these are fewer than on the maculae.

D. Discussion

Werner (1933) defined the striola as a regional reinforcement of the gelatinous substance enveloping the crystals over the maculae (see p. 94). However, he also used the term striola to localize the special region of the sensory epithelium under this area of the gelatinous substance (Werner, 1933, 1940). The term striola, which means little stripe, is used in the present study to denote the sensory epithelium and the statoconial membrane in this central, curved area on the maculae. Its position, as observed in surface specimens, corresponds essentially to Werner's area, reconstructed from serial sections. Noticeable differences are due chiefly to the fact that the shape of the whole sensory epithelium in Werner's illustrations differs slightly from the present observations. This may have been due partly to perspective foreshortening, which, according to Werner (1933), explains why only certain areas of his drawing of the macula utriculi are seen in their natural proportions. It should be stressed, however, that the striola on the macula sacculi, in the present study in the guinea pig, was found to run further posteriorly than Werner (1933) showed.

Lorente de Nó (1926) described a central zone, zone b, with a special innervation in the macula utriculi of the mouse (see p. 72). Its position corresponds essentially to that of the striola. In the posterior regions of the macula utriculi, however, zone b did not turn off in a medial direction, as the striola does in all mammals investigated here. Nomura *et al.* (1965) localized the striola on the macula sacculi of the guinea pig considerably further inferiorly, and not so far anteriorly, as shown in the present investigation.

Werner (1940) concluded that the striola of the guinea pig had large sensory cells. The nuclei of the supporting cells were said, furthermore, to be basal in position, whereas those of the sensory cells were located higher, almost in a single layer (Werner, 1933). Neumann and Neubert (1958), who investigated the macula utriculi in serial sections and surface preparations in the same species, were not able to verify these findings. The present observations, however, are in complete agreement with those of Werner. Furthermore, the sensory cells in the striola had a large free surface and a smaller number of sensory cells per unit surface area was found here than in the peripheral regions of the maculae. These findings were made in all species investigated.

Engström and Wersäll (1958a) were of the opinion that the striola on the macula utriculi had an especially large number of type I cells. Spoendlin (1966) found the same on the macula sacculi. The present observations in the guinea pig show that about two-thirds of all the sensory cells in this area are type I cells and that there is an approximately equal number of the two cell types peripherally. Engström and Wersäll (1958a) also mentioned that type I cells at the apex contained numerous mitochondria. This is particularly true for the large cells in the striola of guinea pigs.

Intraepithelial spaces in the vestibular sensory regions have been described previously. Kolmer (1927) was of the opinion that they were a result of poor fixation. On the other hand, Neumann and Neubert (1958) observed intraepithelial spaces centrally in the macula utriculi of guinea pigs treated with streptomycin, and assumed that the spaces were the result of degeneration of the sensory cells. The present study, however, shows that these spaces are normally present in the striola of the macula utriculi in the guinea pig. They have previously been described by Werner (1933) as cystic spaces. The spaces are most common in the striola of the macula utriculi, but they are also seen in the striola of the macula sacculi and sometimes centrally in the sensory epithelium on the cristae. The spaces do not seem to be related to nerve fibres or to subepithelial blood vessels. Nor do they appear to communicate with the endo- or peri-lymphatic spaces. In the organ of Corti the spaces are said to be filled with a special extracellular fluid called cortilymph (Engström, 1960; Engström *et al.*, 1965), probably resembling perilymph more than endolymph in its ionic composition (Rauch, 1964). It is possible that the intraepithelial spaces in the vestibular sensory epithelium are also filled with a special fluid.

Wersäll (1956) and Spoendlin (1965) observed that type I cells were concentrated on the summit (vertex) of the cristae, in contrast to type II cells, which were mainly found more peripherally. The present studies in the guinea pig did not confirm this. There were also, in the peripheral regions, more of type I than of type II cells. The relationship between type I and type II cells was also approximately the same peripherally as centrally. However, the absolute number of type I cells per unit surface area was larger peripherally, where the density of sensory cells was far greater than in the central regions. The completely marginal zone of the sensory epithelium, however, contained more type II than type I cells. The present investigation, however, is at variance with that of Wersäll (1956), who described a marginal zone of exclusively type II cells.

In the literature the sensory epithelium on the top (summit, vertex) of the crista is generally discussed in contrast to the sensory epithelium at the periphery of the crista. This division, which is used chiefly in describing regional differences in the distribution of type I and type II cells and in the sensitivity to ototoxic antibiotics, etc. of the crista, seems, however, to be inadequate. In view of the structure of the sensory epithelium, and of its reaction to ototoxic antibiotics (Lindeman, 1967, 1969), it seems more correct to divide the sensory epithelium into central and peripheral regions. The central zone, which only occupies part of the summit of the crista, is characterized by scattered sensory cells, most of which have a large free surface, and a smaller number with a small free surface. The periphery of the sensory epithelium, some of which is also on the summit of the crista, is characterized by densely packed sensory cells with a small free surface. The transition between the two regions is gradual.

In the supporting cells, especially in the apical part, there are numerous granules, 0.2—0.3 μ in diameter (Wersäll, 1956) which, according to Wersäll (1967) contain smaller granules. On the basis of these findings Wersäll assumed that the supporting cells had a secretory function. A secretory function has been suggested by a number of authors (Retzius, 1884; Kolmer, 1927; Flock, 1964; Smith, 1967; and others). Dohlman and Ormerod (1960), however, found no sign of secretion of endolymph from the sensory epithelium. Smith (1967) discussed whether the supporting cells could take part in the formation of the gelatinous substance in the statoconial membranes and cupulae.

In the supporting cells of the sensory epithelium, some strongly osmiophilic granules with a diameter of up to 0.5 μ have also been described (Wersäll, 1956; Engström, 1965). The present observations showed that the largest of these granules in the maculae of the guinea pig had a diameter of 2 μ, and in the cristae 3 μ, and that they were characteristically localized in the striola and the central areas of the cristae. Similar granules are seen in Hensen cells in the organ of Corti, where many authors consider them to be lipid droplets. After osmium fixation the granules in the vestibular supporting cells often have a very irregular shape, as is usually seen for lipid droplets (Fawcett, 1966). Outside the outer limits of the sensory epithelium, in certain zones, granules with a similar appearance, but even larger and more closely packed, were seen in light and phase-contrast microscopy (Fig. 22a).

A few collapse figures were observed even in so-called normal sensory epithelium. These probably indicate "spontaneous degeneration" of sensory cells.

Similar collapse figures, but in far larger numbers, were seen in the vestibular sensory regions after treatment of guinea pigs with streptomycin and kanamycin (Lindeman, 1967, 1969). These figures are formed after disintegration of the nuclei and cytoplasm of the sensory cells and collapse of the plasma membrane. The cell then loses contact with the surrounding supporting cells, except in some regions, and this then determines the appearance of the collapse figure (see also Engström *et al.*, 1966).

These "spontaneously degenerated" cells, visible as collapse figures, represented only just over $1^0/_{00}$ of the total number of sensory cells in the guinea pig. Under normal conditions, they thus amount to only a very small proportion of the total number. When assessing the effect of noxious agents, such as ototoxic antibiotics, on the vestibular sensory epithelia, however, their existence should be recalled. This is especially true when there are only a few sensory cells degenerated as a result of noxious agents.

If degeneration of a sensory cell has taken place about a month or more previously, the characteristic collapse figure disappears. This is probably because the surrounding supporting cells completely fill the space left by the previous sensory cell. In the organ of Corti in guinea pigs, where the cell pattern is very regular, and where each sensory cell has its definite position, it is possible to demonstrate the position of the degenerated sensory cell with great certainty even after a long survival period. In the vestibular sensory regions, where the cell pattern generally is far more irregular, it is not so easy to localize such degenerated cells. However, if the cell density in an area is especially low, and the cell pattern particularly irregular, this may be an indication of a disappearance of sensory cells. Nevertheless, it is important to realize that there are areas even in normal sensory epithelium where the cell density is especially low. This is particularly true for the medial, marginal part of the pars interna on the macula utriculi (Fig. 20). A section through this area may therefore give the impression of a marked disappearance of sensory cells.

In order to test the applicability of the microdissection technique to the study of pathological sensory epithelium, an approximately 4-month-old anencephalic foetus was examined. Besides marked cochlear changes (see Bredberg, 1968), a (somewhat) central area on the macula utriculi, outside the striola, was seen where sensory cells were completely absent (unpublished observations). Collapse figures were demonstrated in the sensory epithelium on the cristae. It is unlikely that these changes in the vestibular sensory epithelium could have been seen with ordinary histological techniques. The method was also applied to two Dalmatian dogs with congenital deafness. In addition to extensive cochlear damage (see Bredberg, 1968), pronounced changes of the saccule and the statoconial membrane were seen (unpublished observations). These vestibular changes were observed in the dissecting microscope with low power magnification ($\times 40$). Light and phase-contrast microscopy of the macula sacculi showed marked reduction of sensory cells, collapse figures and a very irregular cell pattern in certain regions.

Counts of sensory cells in the surface specimen can be carried out in three ways: By counting 1. the hair bundles, 2. the free surfaces of the sensory cells or 3. their nuclei. In most areas the nuclei of the sensory cells lie at rather different

levels in the sensory epithelium, and this makes it necessary to focus in different
planes. One can therefore scarcely avoid to count the same nucleus twice, or to
omit counting a nucleus. The first two methods are therefore to be preferred. A
combination of these formed the basis of the present quantitative studies. The
hair bundles are seen best, but since they sometimes loosen during preparation,
the microscope was always focussed also on to the surface of the epithelium, to
avoid missing any sensory cells. Especially in the peripheral regions of the cristae,
where the cells lie very close together, it was a great advantage to have the hair
bundles as identification marks, since sensory cells are less distinctly seen in this
area when the observer focusses on the epithelial surface.

While the maculae, on account of their fairly flat shape, were always well
suitable for surface studies, the sensory epithelia on the cristae were less suitable
when the specimen had been stored in alcohol for a long time. The epithelia, which
follow the saddle shape of the cristae, were then more difficult to flatten under
the cover glass. This gave an impression of a greater density of sensory cells in
the peripheral regions and of a smaller area of sensory epithelium. Sensory
epithelia fixed in 1.5% osmium tetroxide for 3 hours followed by 10 hours in 70%
alcohol appeared, however, to be well suited to these studies. The length of the
surface of the sensory epithelium in such specimens was in agreement with that
of the surface of cristae without sensory epithelium.

The sensory epithelium of the cristae was divided into three regions for the
purpose of quantitative studies on the number of sensory cells, but this division
is artificial. Nevertheless, such a division was regarded as advantageous because
of the marked regional differences in the density of sensory cells. Both the peripheral and the central zone showed a fairly constant density. The density of
sensory cells in the intermediate zone, however, varied considerably. Since it was
difficult to define the limits of this area, it was easy to obtain a misrepresentation
of the central or the peripheral part of this zone, with a subsequent error in
estimation of the mean density.

Previous estimates of the size of the maculae in the guinea pig are, on the
whole, in agreement with the present findings since de Burlet and de Haas (1923)
estimated that the surface area of the macula sacculi was 0.49 mm², and that
of the macula utriculi, 0.48 mm². Werner (1933) reported the area of the macula
utriculi to be between 0.500 and 0.598 mm², and pointed out that the area was
independent of the size of the animal.

Some preliminary results of quantitative investigations have been published
previously (Engström et al., 1966a; Lindeman, 1967). Apart from these publications,
and Mair and Fernandez's (1966), calculations of the number of cells on two
cristae laterales in the cat, there is apparently no information on the number of
sensory cells in the vestibular sensory regions.

It is interesting to correlate numerical data concerning the sensory cells with
the number of nerve fibres in the vestibular sensory regions. Wersäll (1956) found
an average of 1,178 myelinated fibres to the posterior ampulla in the guinea pig.
Gacek and Rasmussen (1961) observed an average of 1,522 fibres to the posterior
ampulla, 1,592 fibres to the lateral ampulla, 1,797 fibres to the anterior ampulla,
1,532 fibres to the macula sacculi and 1,703 fibres to the macula utriculi in the
guinea pig. These figures include both the afferent and the small number of efferent

fibres (see Gacek, 1966). If the present findings regarding the number of sensory cells are compared with Gacek and Rasmussen's data, it is seen that the cristae have a relatively richer innervation than the maculae. The findings show that the former have from 2 to 4 sensory cells for each fibre, the corresponding number on the maculae being from 4 to 6.

Preliminary observations made on one cat, where the area of the maculae was based on measurements of the statoconial membranes, showed that the macula sacculi had an area of 0.779 mm², with the number of sensory cells at about 11,250. The macula utriculi had an area of 1.560 mm², with the number of sensory cells at about 26,350. Gacek and Rasmussen (1961) found 2,089 fibres to the macula sacculi and 2,694 to the macula utriculi in the same animal. This could correspond to 5—6 sensory cells per fibre in the macula sacculi, and to 9—10 sensory cells per fibre in the macula utriculi.

VI. Morphological Polarization of the Sensory Cells

A. Introduction

The adequate stimulus for the vestibular sensory cells is generally considered to be a shearing motion of the cupula and statoconial membranes on the surface of the sensory epithelium. The mechanical stimulus probably acts primarily on the hairs of the sensory cells. Secondarily, the energy is converted into action potentials in the sensory nerve fibres innervating the cells.

The sensory hairs projecting from the free surface of each sensory cell are of differing lengths (Retzius, 1871, 1884). They are, moreover, not identical, but of two types. The difference between the true sensory hairs of a cell and the flagellum has been demonstrated in the past by simple light microscopy, which has also shown that the flagellum emerges from a diplosome in the cuticle-free zone in the peripheral part of the sensory cell (N. van der Stricht, 1908; Held, 1909, 1926; O. van der Stricht, 1918; Kolmer, 1924, 1927). These two types of hairs have been verified with the electron microscope by Wersäll (1956). He observed peripherally in the hair bundle a structure which he called a kinocilium and which was similar to the cilia in the respiratory epithelium. Its axial filamentous complex emerged from a basal body at the free surface of the cell with the same structure as a centriole. The remaining hairs on the sensory cell were similar to stereocilia, and were attached at their roots to the cuticular plate of the cell. Lowenstein and Wersäll (1959) later showed that the hair bundles on cristae ampullares in elasmobranchs were orientated in accordance with a definite pattern. On the lateral crista the kinocilia were nearer the utricle than the associated bundles of stereocilia, while the opposite was true of the anterior crista. A similar difference in orientation of the hair bundles was later shown to exist in mammals (Wersäll, 1961). Lowenstein and Wersäll (1959) showed that this morphological polarization of the sensory epithelium could be correlated with previous electrophysiological observations (see below). The orientation of the hairs on the sensory cells thus seemed to be of decisive importance for the directional sensitivity of the cells.

While the sensory cells on the same crista are polarized in the same direction, conditions on the maculae are more complicated. Engström et al. (1962) observed that the orientation of the hairs was the same over large areas of these regions. This pattern reversed itself along a certain line, in such a way that the kinocilia on opposite sides of the dividing line faced each other. Lowenstein et al. (1964) also showed that the macula sacculi in fishes was divided into two areas with opposite polarization, but they could not find any pattern in the orientation of the hair bundles on the macula utriculi. Flock (1964), on the other hand, concluded that the sensory cells on the macula utriculi of the fish were polarized in accordance with a very definite pattern. He showed that the sensory epithelium was divided into two areas by a curved line, and that the kinocilia on each side of it were nearer to the dividing line than their associated bundles of stereocilia. Spoendlin (1964, 1965) showed a basically similar pattern of polarization on the macula utriculi of guinea pigs, and also that the sensory epithelium even on the macula sacculi was divided in two by a curved line. The kinocilia on the opposing sides were more distal to the dividing line than the associated bundles of stereocilia (Spoendlin, 1965).

B. Material and Methods

Temporal bones from guinea pigs, rabbits, cats, squirrel monkeys and man (foetus and adult) were included in the investigation. The vestibular sensory epithelium was fixed in osmium tetroxide as previously described (see p. 8). Studies of the sensory hairs and their orientation were carried out by light and phase-contrast microscopy, and by electron microscopy.

Surface specimens of the sensory epithelium for light and phase-contrast microscopy were prepared as described on p. 9. For a study of the orientation of the hair bundles it was especially important that the surface preparations were mounted in large quantities of glycerin, otherwise the hairs were easily folded down when the cover glass was put in position. On the other hand, when the length of the sensory hairs was studied the sensory epithelium was mounted in less glycerin, or light pressure was exerted on the cover glass. The hairs were thus pressed down onto the surface of the epithelium and their entire length was revealed.

For a study of the structure and orientation of the hair bundles under the electron microscope, the vestibular sensory regions were embedded in Epon. Sections were made parallel or perpendicular to the epithelial surface.

C. Observations
1. Macula utriculi

One kinocilium and several stereocilia project from the free surface of every sensory cell (Figs. 13, 26a, b, 27a—c). The kinocilium contains nine peripheral double tubular filaments connected with the triplet bundles of the basal body in a cuticle-free area on the free surface of the cell. It also contains two central single tubular filaments which end just above the free surface of the cell. The stereocilia consist of an outer plasma membrane and an inner protoplasmic structure with a central core, which continues down to the cuticular plate and often penetrates it. The stereocilia are arranged in parallel rows in a completely regular pattern (Fig. 27a, b). In the guinea pig there are between 50 and 100 stereocilia on each sensory cell, and in the monkey the average number appears to be higher. The kinocilium is the longest of the hairs (Figs. 26a, 28). In the guinea pig it measured 18—20 μ, in the rabbit about 25 μ. The stereocilia vary

MORPHOLOGICAL POLARIZATION

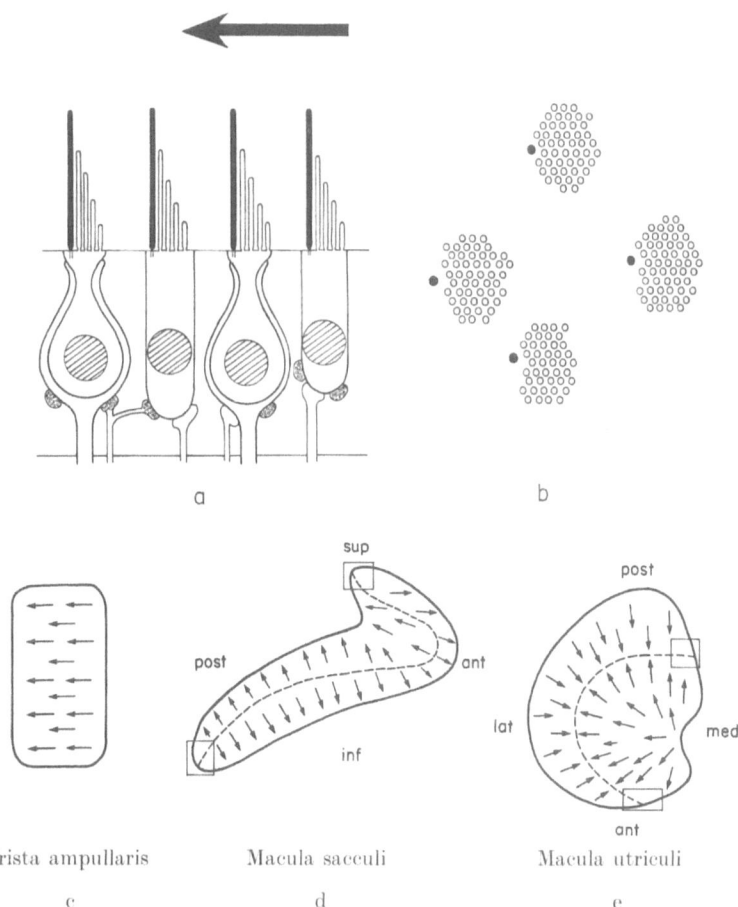

Fig. 26a—e. Diagram illustrating the morphological polarization of the sensory cells and the polarization pattern of the vestibular sensory epithelia. The morphological polarization (arrow) of a sensory cell is determined by the position of the kinocilium in relation to the stereocilia. a Section perpendicular to the epithelium. Note increasing length of stereocilia towards the kinocilium. b Section parallel to the epithelial surface. c The sensory cells on the crista ampullaris are polarized in the same direction. d Macula sacculi and e macula utriculi are divided by an arbitrary curved line into two areas, the pars interna and the pars externa, with opposite morphological polarization. On the macula sacculi the sensory cells are polarized away from the dividing line, on the macula utriculi towards the line. Constant irregularities in the polarization pattern are found in areas corresponding to the continuation of the striola peripherally (rectangles in d and e)

considerably in length. Those nearest to the kinocilium are the longest, the length of the stereocilia gradually diminishing away from the kinocilium (Figs. 26a, 27a—c). In the guinea pig the longest stereocilia measured 12—15 μ, in the rabbit about 14 μ.

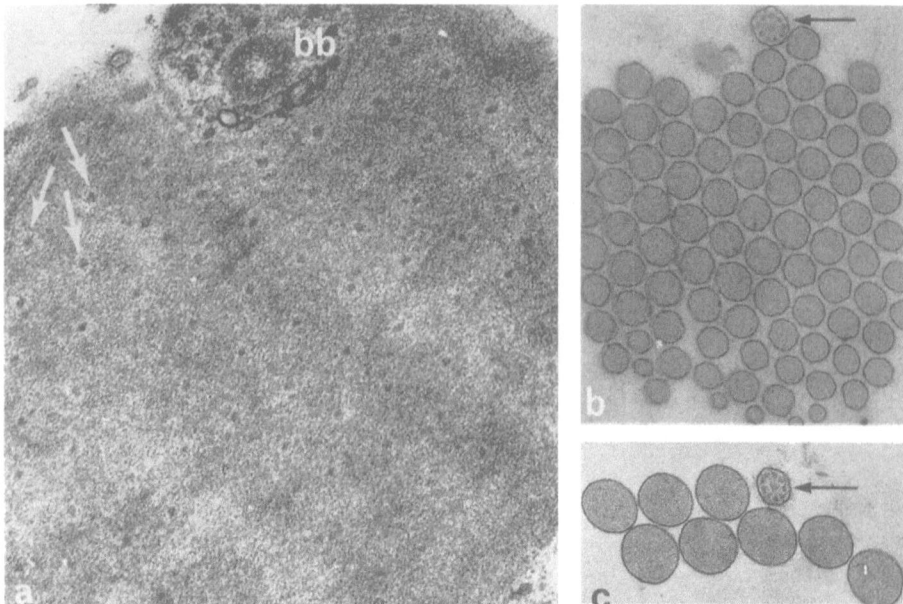

Fig. 27a—c. Electron micrographs from the macula utriculi of the squirrel monkey, illustrating the arrangement of the sensory hairs. A section just below the free surface of the sensory cell (a) reveals the basal body (*bb*) and the regular arrangement of the stereociliar roots (arrows) in the cuticular plate. b shows a section close above the free surface of the sensory cell. Arrow points to kinocilium. c shows more distal section and reveals the kinocilium (arrow) and the longest of the stereocilia. The stereocilia are thicker distally than they are closer to the cell surface. a × 33,000; b × 20,000; c × 17,400

The position of the kinocilium on the free surface of the sensory cell is eccentric. The sensory cell is thus morphologically polarized, and the direction of polarization is determined by the position of the kinocilium in relation to the stereocilia of the same cell (Fig. 26a, b). In surface specimens the kinocilium is usually seen as a darker dot at the periphery of the hair bundle (cp. Fig. 29; arrows), and the morphological polarization of the cell can thus easily be determined. The kinocilium seems to be more flexible than the stereocilia. This means that in surface specimens, the position of the kinocilium in relation to the stereocilia may change on focussing at different levels. The morphological polarization of the sensory cell must therefore be determined by focussing on the hair bundle at the surface of the epithelium.

The sensory cells within small areas are generally polarized in the same direction, with only small deviations from the main direction of orientation. However, some cells are seen with reversed polarization, with the kinocilium on the opposite side of the hair bundle. A few cells are observed with deviations of orientation of between 60—120° from the main direction of polarization. In addition, irregularities in the polarization pattern are seen more often in certain areas of the maculae than in others (see later). Even though the morphological polarization of the cells may thus vary to some extent, this does not alter the main orientation in the area.

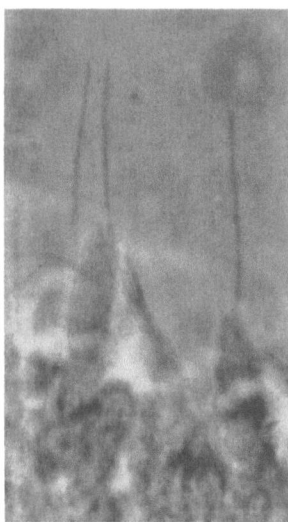

Fig. 28. Photomicrograph showing hair bundles on the macula sacculi of the rabbit. The kinocilium is much longer than the associated stereocilia. × 1,560

The basic features of the polarization pattern of the macula utriculi were the same in all the species investigated. Minor differences are shown in Fig. 30. The macula utriculi is divided into two areas — the pars externa and the pars interna (Fig. 19a) — by an imaginary curved line. The sensory cells in each of these two areas are polarized towards this line (Figs. 26e, 30). The dividing line has a slightly wavy course, through the striola (Fig. 31), and it continues in the long axis of the striola to the periphery of the sensory epithelium (Fig. 19a).

The sensory cells in the pars interna are morphologically polarized in such a way that the directions of polarization diverge like a fan, in posterior, lateral and slightly anterior directions from a medial, marginal area (Figs. 26e, 30). It is in this area that the outer border of the sensory epithelium folds in, and the sensory cells are found to lie very scattered (see p. 43). Constant irregularities in the morphological polarization of the sensory cells are found in those areas which correspond to the continuation peripherally of the striola (rectangles in Fig. 26e).

In the striola, the orientation of the sensory cells seems to be more irregular than in the rest of the sensory epithelium. However, a study of surface specimens also reveals other regional differences. The hairs on those sensory cells in the striola which have a large free surface appear to be thick and club-shaped (cp. Fig. 32a). They often spread out like a fan (cp. Fig. 33), making it difficult to identify the kinocilium and thus to determine the direction of polarization. On those sensory cells which have a small free surface, and on the cells outside the striola, the hairs are thinner and they are usually aggregated into distinct bundles (cp. Figs. 32b, 33). These regional differences in the appearance of the sensory hairs were constant features in the guinea pig, rabbit and cat, but systematic studies were not performed in monkey and man.

As shown in the Fig. 26e, the sensory cells on the macula utriculi are orientated in anterior, posterior, medial and lateral directions. The numbers of sensory cells

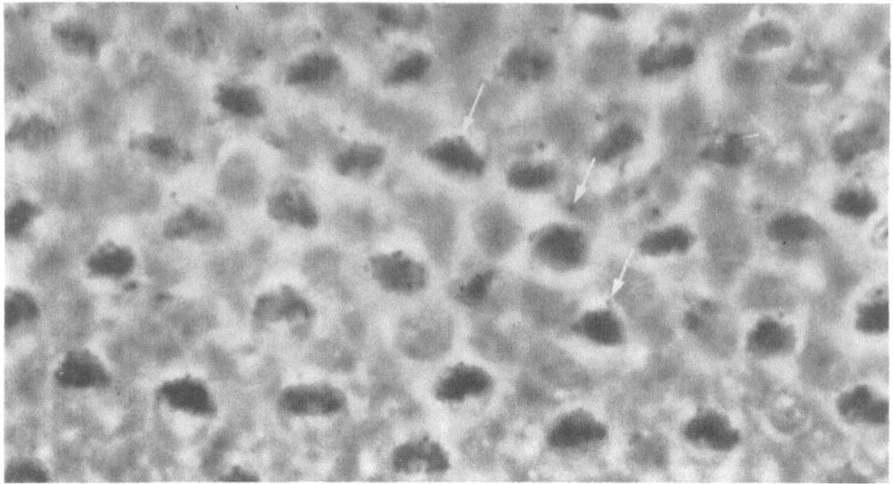

Fig. 29. Surface preparation showing hair bundles in the periphery of the crista in the guinea pig. The kinocilia are seen as dark dots (arrows) at the same side of the hair bundles. × 1,560

in the two areas of opposite polarization, i.e. the pars interna and pars externa, have been studied in the guinea pig, in which they have been found to be more or less the same (see p. 43).

2. Macula sacculi

There is no difference between the macula utriculi and the macula sacculi regarding the ultrastructure of the kinocilia and stereocilia, the length of the hairs or differences in the appearance of the hairs between the striola and the peripheral regions (Figs. 32a, b, 33).

The macula sacculi can also be divided into two areas — the pars externa and the pars interna — by an imaginary curved line. The sensory cells here, however, are morphologically polarized away from this line (Figs. 26d, 30). The dividing line runs along the middle of the striola. Fig. 30 shows the polarization pattern in the guinea pig, rabbit, cat, monkey and man. As in the macula utriculi, so in the macula sacculi there are also constantly present in those areas which correspond to the peripheral continuation of the striola, irregularities in the orientation of the sensory cells (rectangles in Fig. 26d).

In the rabbit and cat there is a narrow antero-inferior marginal zone where the sensory cells are orientated in a direction opposite to the rest of the pars externa (Fig. 30). The dividing line of polarization here has a very irregular wavy course at the outer limits of the sensory epithelium. No such marginal zone was found in the guinea pig, monkey or man. On the macula sacculi the sensory cells are polarized chiefly in two directions (Fig. 26d). In the guinea pig the number of sensory cells orientated in an antero-inferior direction slightly exceeds the number of sensory cells orientated in a postero-superior direction.

3. Sensory Epithelia on the Cristae ampullares

The sensory hairs on the cristae ampullares are longer than those on the maculae, but their ultrastructure is the same. As on the maculae, so on the cristae

POLARIZATION PATTERN

MACULA SACCULI   MACULA UTRICULI

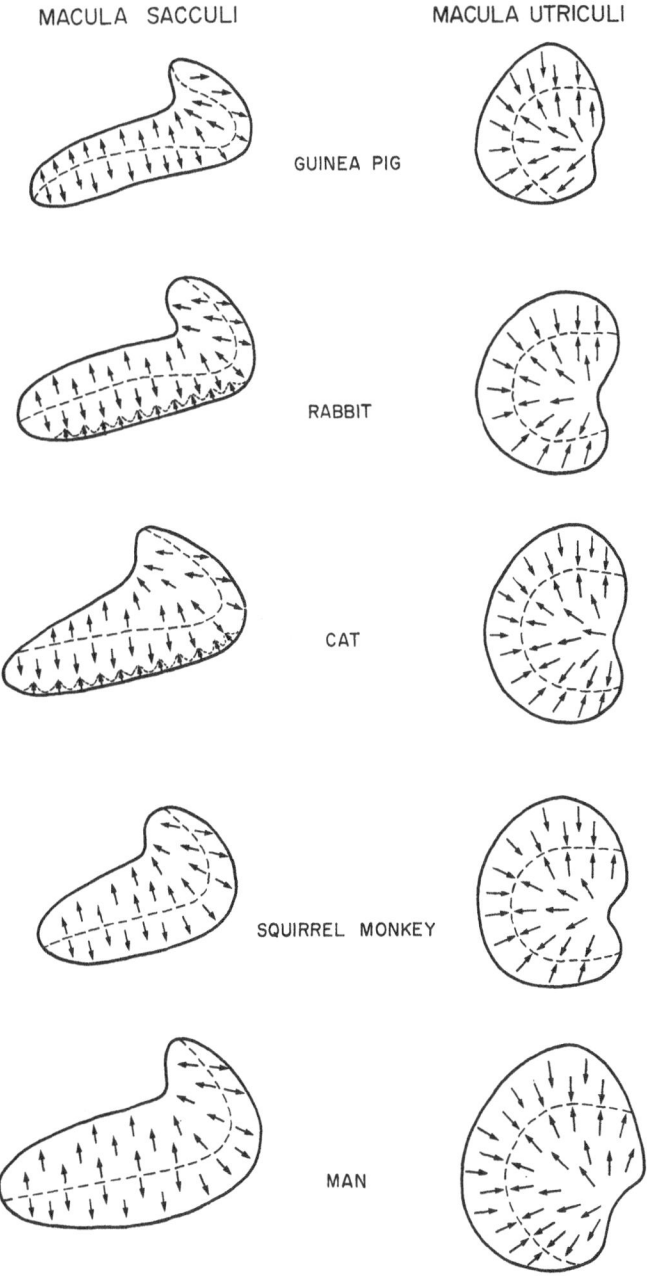

GUINEA PIG

RABBIT

CAT

SQUIRREL MONKEY

MAN

Fig. 30. Diagram illustrating the polarization pattern in different species of mammals. In the macula sacculi of the rabbit and cat, a marginal, antero-inferiorly located zone contains sensory cells, morphologically polarized in a direction opposite to that of the sensory cells in the remaining part of the pars externa

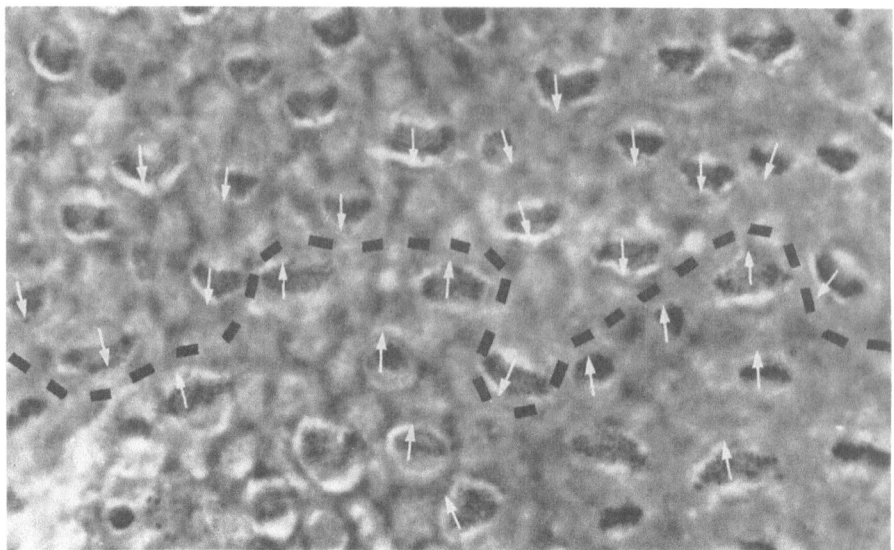

Fig. 31. Surface preparation from the macula utriculi in the guinea pig. The arbitrary line (broken line), dividing the macula into two areas, lies in the middle of the striola. It has a slightly wavy configuration. The direction of polarization is indicated by arrows. × 1,560

the kinocilium is longer than the stereocilia, which gradually decrease in length as their distance from the kinocilium increases. In surface specimens, obvious regional differences in the length and appearance of the hairs are seen. In the peripheral areas the stereocilia are thin and slender, and the kinocilia often seem to end distally in a bulbous thickening (Fig. 34a). In the guinea pig, the longest kinocilia at the periphery of the crista were about 80 μ in length. The maximal length of the stereocilia, however, varied considerably from crista to crista. The longest stereocilia were sometimes about 45 μ, more often about 60 μ.

The sensory hairs are shorter in the central than in the peripheral areas of the cristae. The stereocilia in the central areas also appear to be slightly thicker, and the distal part is often club-shaped (Fig. 34b). This is most noticeable in sensory cells with a large free surface. These regional differences in the length and appearance of the hairs were seen in the guinea pig, rabbit and cat.

The morphological polarization of the sensory cells may be easily determined in surface preparations from the cristae ampullares (Fig. 29). The sensory cells on any single crista are orientated in the same direction (Fig. 26c). The sensory cells on the lateral crista are polarized towards the utricle, those on the anterior and posterior cristae away from the utricle.

D. Discussion

By recording action potentials from single vestibular fibres from the cristae (Lowenstein and Sand, 1940; Adrian, 1943), and from the maculae (Lowenstein and Roberts, 1950), it was shown that even when the head was not moved, impulses travelled in a central direction, and that this spontaneous activity was

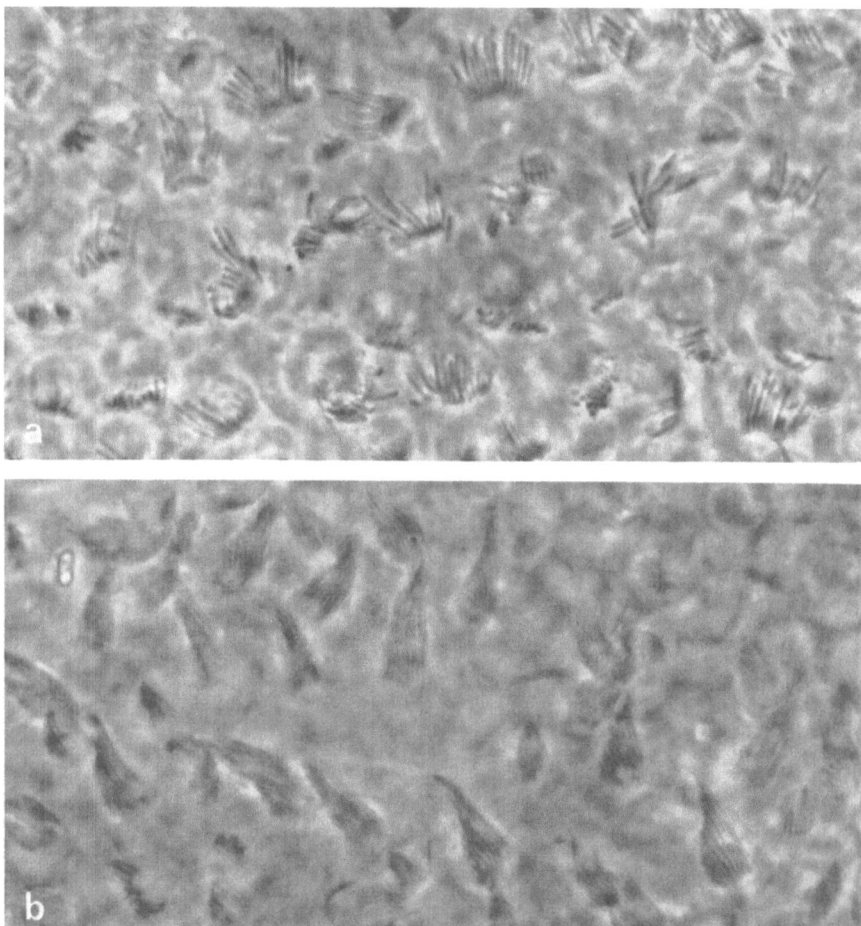

Fig. 32a and b. Surface preparations from the macula sacculi in the guinea pig. In the striola (a) the stereocilia are thicker and more club-shaped than in the peripheral regions (b). × 1,560

present in the majority of nerve fibres. Displacement of the cupula or statoconial membrane altered this activity in the nerve fibres. Displacement in an utriculo-petal direction (i.e. towards the utricle) of the cupula on the lateral crista was accompanied by depolarization of the sensory epithelium (Trincker, 1957) and increase of impulses in the nerve fibres (Lowenstein and Sand, 1940; see also Gernandt, 1949). On the other hand, utriculofugal displacement of the cupula (i.e. away from the utricle) produced hyperpolarization of the sensory epithelium and decrease of activity in the nerve fibres. The reverse conditions were found in the vertical semicircular ducts. In recordings from the macula utriculi (Trincker, 1959), displacement of the statoconial membrane in one direction produced hyper-polarization; displacement in the opposite direction, depolarization.

These findings showed that the vestibular sensory epithelium was functionally polarized. It seemed that a morphological correlate to this functional polarization

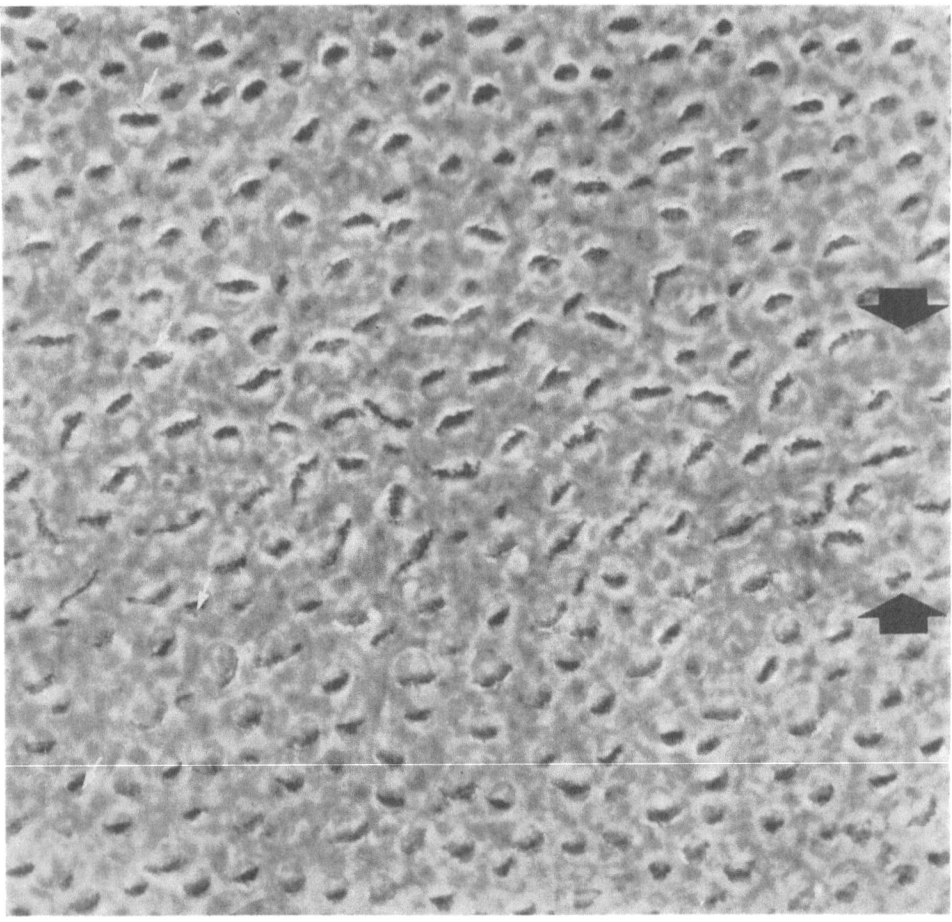

Fig. 33. Surface preparations from the macula sacculi of the guinea pig. The bundles of sensory hairs have a different appearance in the striola (between thick arrows) from that on either side of the striola. White arrows indicate kinocilia. × 780

of the sensory epithelium had been found when Lowenstein and Wersäll (1959) established that the hair bundles on the lateral crista were orientated towards the utricle, and those on the anterior crista away from the utricle. Stimulation of the sensory hairs in a direction away from the stereocilia towards the kinocilium was thus considered to give depolarization of the sensory cell and an increase in the frequency of impulses in the nerve fibres; whilst stimulation of the sensory hairs in the opposite direction was considered to result in hyperpolarization and a reduction in the frequency of impulses in the nerve fibres (Flock and Wersäll, 1962, 1963; Dijkgraaf, 1963; and others). Furthermore, it was assumed that the effect of any force was proportional to the size of the component in the direction of polarization of the sensory cell (Flock, 1965).

It has also been possible to correlate functional and morphological polarization in the lateral line organ and in the organ of Corti. The sensory cells in the lateral

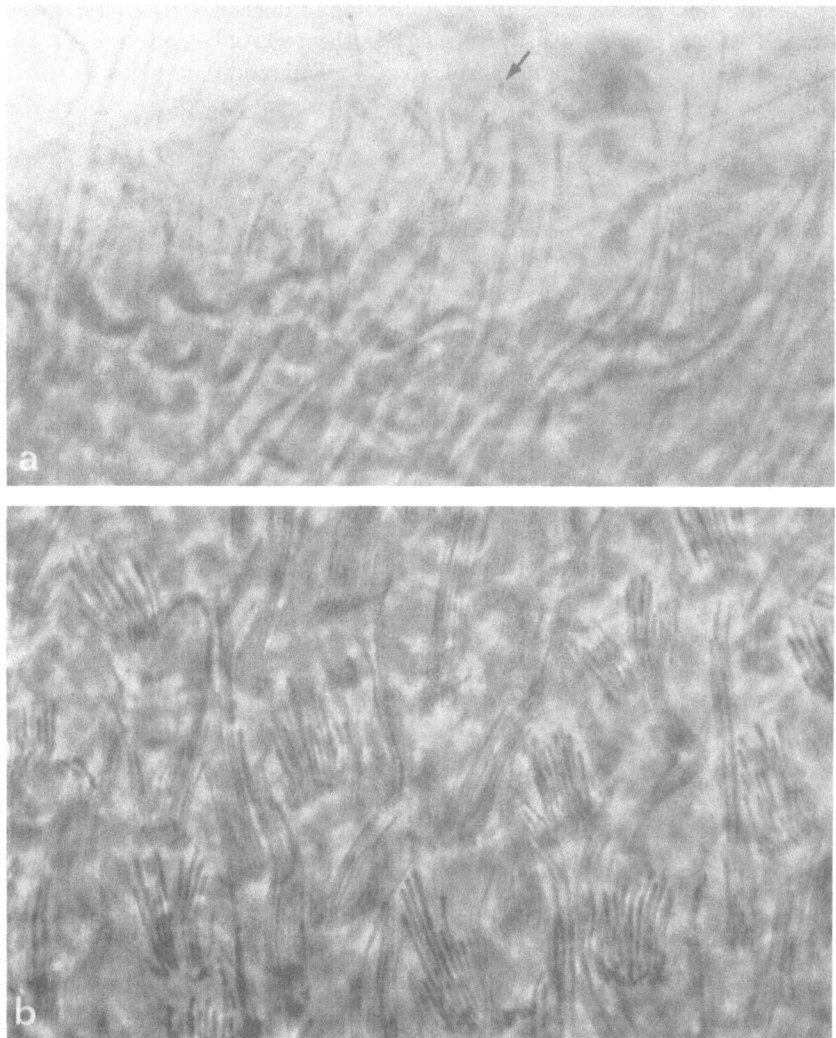

Fig. 34a and b. Photomicrographs of hair bundles from the crista in the guinea pig. In the central regions (b) the hairs are thicker, more club-shaped and also shorter than at the periphery (a). Note thickening of the most distal end of the kinocilia (arrow in a).
a and b × 1,560

line organ are orientated in the direction of the long axis of the canal. Sensory cells polarized towards the head alternate with sensory cells polarized towards the tail (Flock and Wersäll, 1962; Dijkgraaf, 1963, Flock, 1965). One group of cells is activated by a flow of fluid in one direction, the other by a flow in the opposite direction (Flock and Wersäll, 1962; Dijkgraaf, 1963; Görner, 1963; Flock, 1965).

The hair cells in the organ of Corti lack a kinocilium, but they have a basal body (centriole), nearer to the stria vascularis than the associated stereocilia (Engström *et al.*, 1962; Flock *et al.*, 1962). According to Békésy (1953) the hair cells are stimulated by a tangential displacement of the tectorial membrane in a radial direction. Thus functional and morphological polarization may also be correlated in the organ of Corti (Engström *et al.*, 1962; Flock *et al.*, 1962; Flock and Wersäll, 1963).

Other factors, besides the eccentric position of the kinocilium and basal body, have also been mentioned as a cause of directional sensitivity in the cells. The gradual increase in length of the stereocilia, with the longest stereocilia nearest to the kinocilium, has been discussed by Engström *et al.* (1962) and Ades and Engström (1965). Furthermore, ultrastructural studies have revealed an asymmetrical arrangement of the filaments of the kinocilium and the presence of a basal foot on that side of the basal body which faces away from the stereocilia (Flock, 1964; Lowenstein *et al.*, 1964; Flock and Duvall, 1965; Wersäll and Flock, 1965).

The influence on the sensory hairs of a component acting tangentially to the surface of the sensory epithelium (shearing force), results in the production of a receptor potential in the sensory cells. This receptor potential is considered to be proportional to the degree of displacement of the cupula (Flock, 1965, 1966; see also Trincker, 1965). However, the mechanism of the transfer of energy to the sensory cells it not yet known. Under the electron microscope the stereocilia give the impression of being stiff hairs. It is assumed that these hairs act as microlevers transferring energy to the cuticular plate and the organelles in the upper part of the cell (Engström *et al.*, 1962; Lowenstein and Osborne, 1964; Ades and Engström, 1965; Flock, 1965; Wersäll *et al.*, 1966, 1967).

Engström *et al.* (1962) and Ades and Engström (1965) regarded the kinocilium and basal body respectively as the essential excitable structure in stimulation of the sensory cell, and drew attention to the presence of these structures in other sense organs such as the retina and olfactory region (see Fawcett, 1961) and in the central nervous system (Dahl, 1963). It has also been suggested that the kinocilium in the vestibular sensory cells might act as a motile structure in reverse, by responding to passive deformation with the initiation of an electrical change (Lowenstein and Wersäll, 1959; Lowenstein *et al.*, 1964; Lowenstein, 1966). Thus any displacement of the kinocilium perpendicular to a line through its two central filaments towards the basal foot, should stimulate the sensory cell. This direction corresponds to the direction of the working stroke of a motile kinocilium (Gibbons, 1961). It should, however, be stressed that kinocilia and modified kinocilia are structures that have been observed to an increasing extent in different organs. In the vestibular labyrinth, rudimentary kinocilia have been observed on supporting cells of the sensory epithelium in the ray fish (Lowenstein *et al.*, 1964) and squirrel monkey (Spoendlin, 1966), and in the epithelial cells of the semicircular ducts in man (Iurato, 1967). The present study has shown, furthermore, that in the cat and human foetus all supporting cells in the vestibular sensory epithelia appear to have a centriole or modified kinocilium at the free surface (p. 39).

From what has been discussed above, it seems obvious that it is not known at present how energy is transferred to the vestibular sensory cells or how the mechanoreceptors in the inner ear convert the mechanical energy to nerve impulses. Nor is it known whether it is the eccentric position of the kinocilium and the basal body or the gradually decreasing length of the stereocilia that is responsible for the directional sensitivity of the cells. Further studies at the ultra-structural level should lead to a better understanding of this problem.

It seems, however, to be a general principle that the functional polarization (directional sensitivity) of sensory cells coincides with their morphological polarization. Thus, a record of the morphological polarization of a sensory region would show also the functional polarization of the same area.

Previous determinations of the orientation of the sensory cells on the maculae and cristae have been carried out on serial sections by light/phase-contrast and electron microscopy, with subsequent reconstruction of the polarization pattern. However, in addition to being simple and rapid, the technique used in the present study, by which surface specimens are obtained, makes it possible to obtain a more accurate picture of the orientation of cells in the different areas. It is possible with this technique to relate the orientation of each sensory cell to the striola or to any other desired area, and to make quantitative estimations of the number of sensory cells which are polarized in a given direction.

The kinocilium seems to be more flexible than the stereocilia. Distal to the epithelial surface it thus often alters its position relative to the stereocilia, an important fact to remember when assessing from sections the direction of polarization. In serial sections it is only possible, therefore, to determine the morphological polarization of a cell when the section has been made right down at the surface of the epithelium. In a surface specimen it is possible to identify the kinocilium wherever it may be over the free surface of the cell, and by focussing down to the cell surface the exact orientation of the cell may be easily determined.

The present observations on the morphological polarization of the sensory epithelium on the cristae ampullares are in agreement with previous findings (Lowenstein and Wersäll, 1959; Wersäll, 1961; Spoendlin, 1965), and they agree largely with those of Spoendlin (1965) concerning the structure of the maculae. However, on the macula utriculi Spoendlin localized the dividing line in the guinea pig considerably more laterally than has been done in the present study. There is also some discrepancy concerning the position of the dividing line in the antero-superior and completely posterior part of the macula sacculi. The present study shows, furthermore, that the orientation of the sensory cells in the rabbit, cat, monkey and man have basically the same polarization pattern as in the guinea pig, and that the dividing line in all these species lies in the middle of the striola. In addition, the macula sacculi has, in the rabbit and cat, a narrow, irregular, antero-inferior marginal zone with a polarization opposite to that of the rest of the pars externa.

The sensory cells in the macula utriculi are orientated in anterior, posterior, medial and lateral directions, and also in the anterior part, in superior and inferior directions. Therefore, any displacement of the statoconial membrane tangentially to the surface of the sensory epithelium will always give an excitatory stimulus to some groups of cells, an inhibitory one to others. It also seems clear that the pattern of impulses in the afferent nerve fibres from the macula utriculi must be very complicated. It is not surprising, therefore, that electrophysiological studies on the function of the maculae have given contradictory results, as pointed out by Flock (1964) (see also Lowenstein, 1966).

With the head in the normal anatomical position, the macula sacculi has an almost vertical position (see p. 15), and the sensory cells are mostly orientated in a postero-superior and an antero-inferior direction. Assuming that the sensory cells show minimal adaptation (see Trincker, 1965), and that the morphological polarization can be correlated with a functional polarization, then there should be, with the head in the normal anatomical position, a fairly constant, relatively high frequency of impulses in the nerve fibres from the pars externa, where the direction of polarization coincides to some extent with the direction of gravity. At the same time one would expect inhibition of spontaneous activity in the fibres from the pars interna.

The present investigation has also given new details concerning the length and location of the sensory hairs and the structure of the hair bundles on the sensory cells. The findings confirm the previous electron microscopical observation that the kinocilium in many cells appears to be located centrally and to be surrounded by stereocilia (Spoendlin, 1965). However, in sections made more basally through the upper region of the sensory cell, a similar centrally-placed kinocilium or basal body was never observed. Therefore the apparently central position of the kinocilium in the hair bundle may possibly result from displacement of the kinocilium during the preparative procedures. This assumption is further supported by the observation that the centrally-located kinocilium is usually surrounded by very irregularly arranged stereocilia.

Flock (1964) demonstrated in fish hair bundles which differed in structure from the normal. He described a "horseshoe type", and showed cells with two kinocilia in which either both of them were found on the same side of the bundle of stereocilia, or one of them was found on each side of the hair bundle.

The regional differences in the appearance of the sensory hairs observed in the present investigation in the guinea pig, rabbit and cat are remarkable. The thick club-shaped hairs in the striola and the central areas of the crista were seen most distinctly in temporal bones which had been stored in alcohol for some days after fixation. It is possible that they may sometimes represent a clumping of several hairs. However, obvious regional differences in the appearance of the hairs were demonstrated in all preparations. Further study of the structure of the sensory hairs with the electron microscope should therefore afford important information.

Reports on the length of the hairs vary. Retzius (1884) found the longest hairs on the maculae to be about 18 μ in the rabbit, 30 μ in the cat and 20—25 μ in man. Iurato (1962) estimated the length of the hairs on the maculae to be about 6—7 μ, while Spoendlin (1965) observed stereocilia between 1 and 12 μ in length. The present investigation shows that the length of the kinocilia on the maculae in the guinea pig is between 18—20 μ, and of the stereocilia between 12—15 μ, while the longest kinocilia in the rabbit are about 25 μ, the longest stereocilia about 14 μ.

The hairs are longer on the cristae. Retzius (1884) reported that they were 30 μ long in the rabbit, 66 μ in the cat and about 28 μ in man, but he was of the opinion that the hairs of undamaged sensory epithelium in man were longer.

Wersäll (1956) found the longest hairs in the guinea pig to measure about 40 μ. Iurato (1962) gave the numbers as 20—25 μ, and Engström et al. (1962) measured them at 40—75 μ. The present study shows that there were great variations in the maximal length of the hairs from one animal to another. The longest kinocilia were in the guinea pig about 80 μ, but the maximal length of the kinocilia on the cristae was generally 45—60 μ, the longest stereocilia being slightly shorter. The investigation also showed that there were conspicuous regional differences in the length of the hairs within the same sensory region, and that the hairs on the crista were longer peripherally than centrally.

VII. Innervation of the Vestibular Sensory Epithelia

A. Introduction

The relation between the nerve fibres and the sensory cells on the maculae and cristae has been subject to intense research since the last half of the 19th century, and perusal of the relevant literature shows that there are very great differences of opinion concerning the innervation of the sensory epithelium. This is especially noticeable in the earliest papers, which show clearly how greatly technique can influence the interpretation of histological findings. Many of the fixing and staining methods gave incomplete impregnation of the nervous elements, and this was obviously the reason for most of the discrepancies.

Retzius was deeply engaged in the problems concerning innervation of the vestibular sensory epithelium (see i.a. Retzius, 1871, 1881a, b, 1884, 1892, 1894, 1905). He described fibres of different thickness which entered the sensory epithelium after losing the myelin sheath. Some fibres passed directly to the base of the hair cells, where they divided into fibrils lying very close to and like cups surrounding the lower part of the sensory cells. Each fibre made contact with 1—5 cells. Other fibres turned horizontally, the branches forming a basal plexus in the sensory epithelium before reaching the hair cells. Kaiser (1891), however, studying specimens fixed in osmic acid, concluded that the axis cylinder did not break up into individual fibrils, but spread out to form a nerve chalice of hyalin ground substance, almost completely surrounding the sensory cell. Niemack (1892/93), who used Ehrlich's methylene blue method, confirmed Kaiser's findings, but he was not able to determine whether the nerve fibres which ended as chalices also anastomosed with the intraepithelial plexus. When the nerve fibres in the plexus branched, swellings were sometimes seen at the sites of division.

Kaiser's and Niemack's findings were later criticized by Lenhossek (1894), Held (1902) and Retzius (1905). Kolmer (1904), who used different silver impregnation techniques, concluded that the nerve fibres split into fibrils which ended in the protoplasm of the sensory cells, a view in keeping with the findings of Bielschowsky and Brühl (1908), and Andrzejewski (1955). Kolmer (1924) therefore regarded the hair cells as true nerve cells. On the other hand, Cajal (1903, 1908, 1909—1911), Lorente de Nó (1926) and Poljak (1927b) observed a separation of the nerve fibres from the protoplasm of the sensory cells, but they agreed with Retzius that the chalices were composed of fibrils.

The above mentioned fundamental problems were not solved until the advent of the electron microscope. Wersäll (1956) showed that the sensory cells of the cristae ampullaris were of two types — which were called types I and II. The innervation of these cells differed. Type I cells were almost completely surrounded by nerve chalices, formed from the axon of thick and medium-sized nerve fibres. The nerve fibres to type II cells were thin or medium-sized and ended in bud-shaped nerve endings on the cells. Other studies showed that the basic features of innervation of the maculi utriculi were the same (Smith, 1956). In addition to the two fundamentally different cell types, Engström (1961) described cells of an intermediate type, surrounded by a half nerve chalice and also innervated by bud-shaped nerve endings. He also observed an overlapping in innervation, so that one and the same fibre might end as a nerve chalice on a type I cell and also give off branches to innervate a type II cell.

Retzius was the first to reveal that fibres of different calibre went to the vestibular sensory regions, but Niemack (1892/93) showed that there were regional differences in the innervation of the maculae and cristae, especially in the frog. Later Cajal (1903, 1908, 1909—1911) in bird embryos observed thick fibres going to the central area of the cristae where they formed chalices, whilst the periphery was supplied by thin fibres which branched to form an intraepithelial plexus. Using Cajal's technique in his studies of the mouse and cat, Lorente de Nó (1926, 1931) was able to confirm Cajal's findings and to make a number of new observations. The central regions of the cristae, a central area of the macula utriculi (corresponding in position essentially to the striola, see p. 53) and the anterior part of the macula sacculi were all innervated by especially thick fibres. He also observed obvious regional differences in the distribution of medium-sized and thin fibres and in the extent of the intraepithelial plexus. Poljak (1927 b), using different techniques in rat and man, was able to verify many of these findings, but he did not find the sharply defined boundaries between different regions that were indicated by Lorente de Nó.

By studying silver impregnated sections and specimens treated with osmium tetroxide, Wersäll (1956) showed in the guinea pig that thick nerve fibres, with a diameter of 6—9 μ continued to the top of the crista, and that each fibre gave rise to 3—4 chalices. On the other hand, medium-sized fibres, 3—5 μ in diameter were distributed over the entire crista and gave off several intraepithelial branches. Some of these formed chalices, but others, together with the thinnest fibres of 1—2 μ, formed a richly ramifying plexus on the sides of the crista considered to supply type II cells. Wersäll (1956), furthermore, differentiated two types of nerve endings to type II cells, some with few and others with many vesicles. Engström (1958) showed that these vesiculated nerve endings were not only in contact with type II cells, but also with the nerve chalices of type I cells. They had the characteristics of presynaptic structures, and Engström suggested that they represented efferent boutons, in contrast to the nerve endings with few vesicles which were considered to be afferent.

It has been possible, experimentally, to follow efferent myelinated fibres in the vestibular nerve (Leidler, 1914; Petroff, 1955; Rasmussen and Gacek, 1958; Gacek, 1960, 1966). Results obtained by histochemical methods have also provided indications of an efferent innervation of the sensory epithelium (Dohlman et al.,

1958; Dohlman, 1960; Rossi, 1961; Rossi and Cortesina, 1962, 1963, 1965a, b; Ireland and Farkashidy, 1961; Hilding and Wersäll, 1962; Nomura *et al.*, 1965; Gacek *et al.*, 1965; Ishii *et al.*, 1967).

It has also been mentioned that the vestibular labyrinth receives autonomic fibres. For example, these have been described outside the sensory epithelium in the walls of the membranous labyrinth (Cajal, 1893; Benjamins, 1925; Palumbi, 1954; Andrzejewski, 1955, 1956) and also in the peripheral branches of the vestibular nerve (Ernyei, 1935; Andrzejewski, 1955). Wersäll (1956) observed in these branches unmyelinated fibres of 0.3—1 μ thickness under the electron microscope, and discussed whether they might be vegetative (Wersäll, 1960). The fibres could correspond to the adrenergic fibres demonstrated with Falck and Hillarp's method (Spoendlin, 1965; Spoendlin and Lichtensteiger, 1966).

B. Material and Methods

Guinea pigs were used in these investigations. The innervation of the sensory epithelium was studied in specimens stained with a modification of the technique *described by Maillet* (1963) (see p. 9). After dehydration, the sensory epithelium was dissected free in xylol, mounted in Canada balsam, and studied by light and phase-contrast microscopy. After successful impregnation, intraepithelial nerve fibres were stained distinctly black.

Some specimens were embedded in Epon after dehydration and sectioned for light and electron microscopy. In most cases the course and distribution of the subepithelial myelinated nerve fibres was studied directly on surface specimens. The terminal parts of the myelin sheaths, just beneath the sensory epithelium, were visible as dark rings under the light microscope. In specimens where both the sensory epithelium and the subepithelial tissue were intact, fibres of different diameters could be related to given regions of the sensory epithelium.

The distribution of the myelinated fibres was also studied in sections from embedded specimens, some of the sections being vertical and some parallel to the surface of the epithelium.

It was easiest to study the innervation of the sensory epithelium on the crista ampullaris. The innervation of this structure will therefore be described first.

C. Observations

1. Innervation of the Sensory Epithelium on the Cristae ampullares

The ampullary nerves, which innervate the sensory epithelium on the cristae, contain both myelinated and unmyelinated fibres (Fig. 35a). The myelinated fibres are of different calibres, and can be classified arbitrarily as thick, medium and thin. The thick fibres run to the central regions of the crista, which is also innervated by a smaller number of thin myelinated fibres. The peripheral areas are supplied by medium-sized and thin fibres. The density of the myelinated fibres is greater in the peripheral than in the central areas.

Shortly before, or at, the point of penetration of the basement membrane, the fibres lose their myelin sheath (Figs. 13, 35b). Occasionally branching occurs at this level, but the majority of the fibres branch intraepithelially. The thick and medium-sized fibres branch once or several times (Fig. 36a), the thickness of the fibres decreasing as the collaterals leave. Intraepithelially the fibres run in all directions, often crossing each other (Fig. 36a). The thick fibres to the central areas innervate in the form of chalices a small number of type I cells. Type I cells innervated by one and the same thick fibre lie close together, but are not arranged

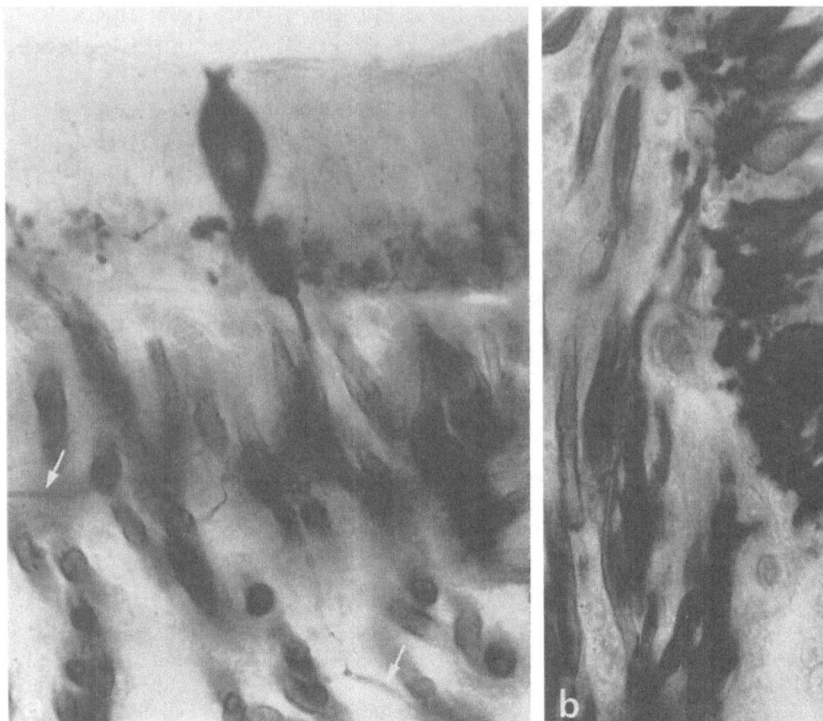

Fig. 35a and b. Sections through the crista ampullaris of the guinea pig. In a both myelinated
and unmyelinated (arrows) fibres are seen below the sensory epithelium. The myelinated
fibres lose their myelin sheath before entering the epithelium. Thick and medium sized fibres
innervate the type I cells as chalices (b). Osmic acid zinc-iodide stain. a × 780, b × 780

in any definite pattern. The medium-sized fibres entering the peripheral regions
may have a long intraepithelial course, and often innervate a large number of
type I cells. In one specimen 10 type I cells were supplied by the same fibre.
Fibres to type I cells also gave branches to type II cells. This innervation of
type I cells is obviously afferent.

The afferent innervation of type II cells seems to be effected by medium-sized
and thin fibres. These make contact with the plasma membrane of the sensory
cells mainly at their bases, by means of bud-shaped sparsely vesiculated nerve
endings. The present author, however, was not able to reveal the intraepithelial
pattern of branching of these fibres.

As mentioned previously, unmyelinated fibres can be seen subepithelially between the myelinated fibres. These are very thin and often do not seem to be
related to the myelinated fibres (Fig. 35a) or to the subepithelial blood vessels.
Some of them penetrate the basement membrane, enter the sensory epithelium,
and disappear in a rich intraepithelial plexus.

The intraepithelial plexus is composed of fine fibres which give off numerous
collaterals and form a basal network in the sensory epithelium, often in several
layers (Figs. 36b, 37). In surface specimens, it is possible to follow these thin
fibres from the utricular to the canalicular side of the sensory epithelium.

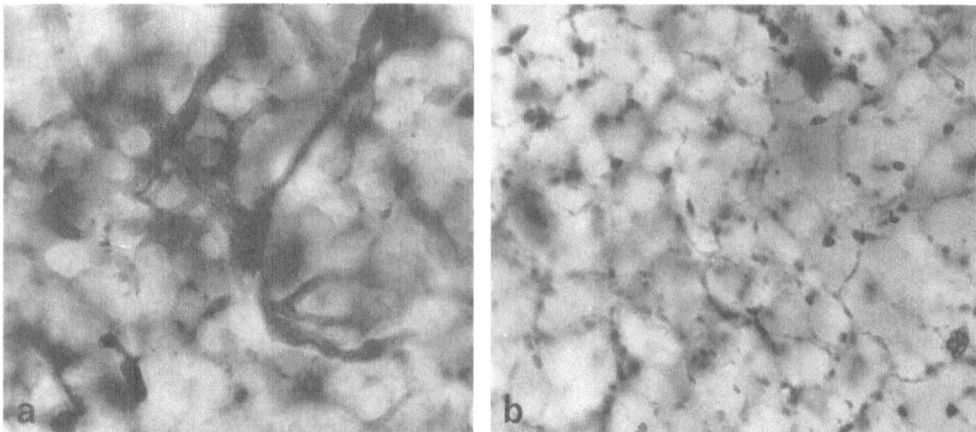

Fig. 36a and b. Surface preparations from the sensory epithelium of the crista in the guinea pig. In a thick fibres, running intraepithelially in all directions and crossing each other, are seen. b shows the intraepithelial plexus of very thin fibres, provided with numerous beads. a and b × 780

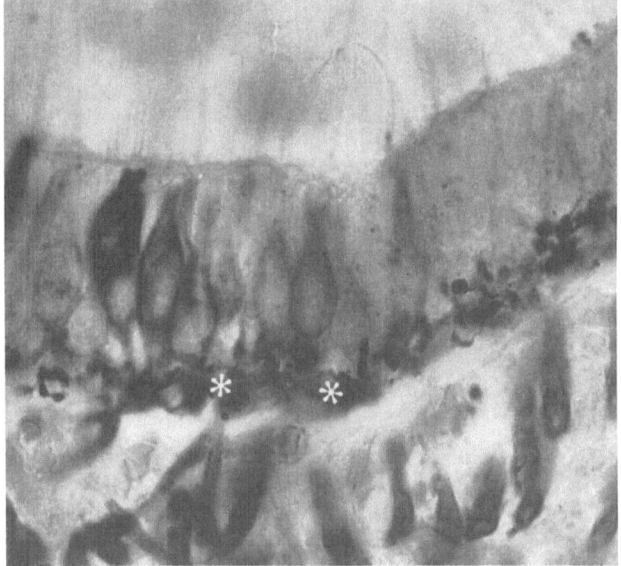

Fig. 37. Section through the periphery of the epithelium of the crista in the guinea pig. * Indicates the plexus of very thin fibres in the basal part of the sensory epithelium. × 780

The fibres in the intraepithelial plexus frequently show bead-like round or ovoid swellings of varying size and shape (Fig. 36b). The swellings are often seen at the points of branching of the fibres, but may also be seen elsewhere. The thin fibres generally run a separate course but may accompany the thicker fibres for short distances. Where this occurs, they twine around the thicker intraepithelial fibres and lie close to them at the swellings. The thin fibres in this intraepithelial

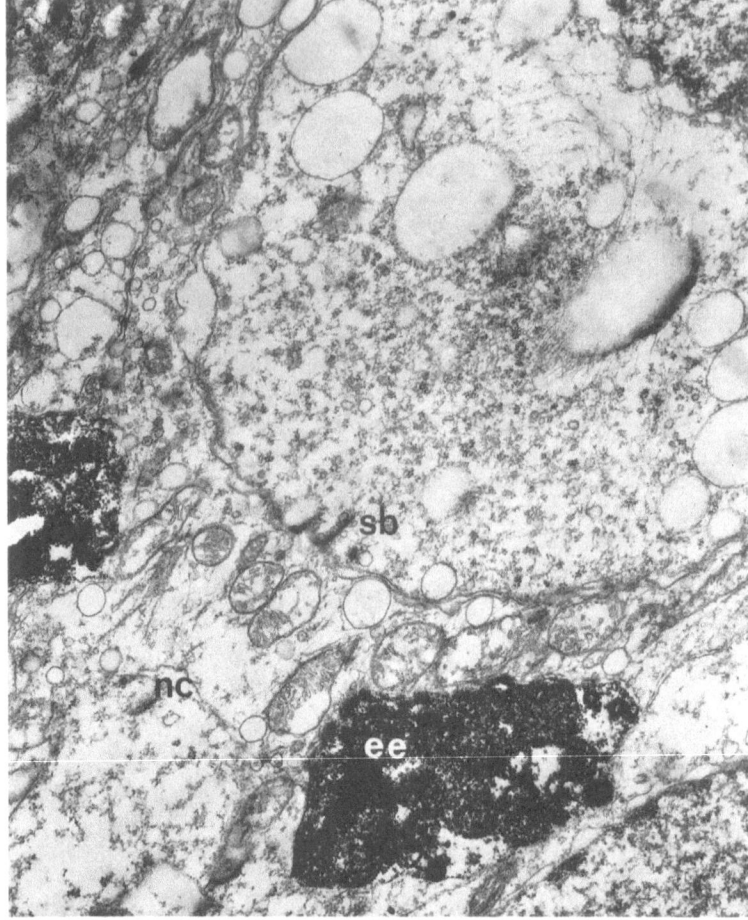

Fig. 38. Electron micrograph from the sensory epithelium of the crista ampullaris in the guinea pig. Osmic acid-zinc-iodide stain. The richly vesiculated nerve endings (*ee*) are heavily stained. *n* nucleus of type I cell, *nc* nerve chalice, *sb* synaptic bar. × 16,000

plexus also twine around the type II cells and the nerve chalices of the type I cells and appear to make contact with these also at the swellings. This was further verified under the electron microscope. Fig. 38 shows two such swellings in contact with the nerve chalice around a type I cell. These may well correspond to the vesiculated nerve endings seen under the electron microscope. In this example one single thin fibre has a large number of swellings which appear to make contact with numerous type II cells, nerve chalices and thicker fibres over a considerable area of the sensory epithelium. Only a minority of the swellings are terminal.

The intraepithelial plexus extends over the entire sensory epithelium on the crista, but regional differences are seen. It is better developed, with more branches and swellings, in the peripheral regions than in the central regions of the sensory epithelium.

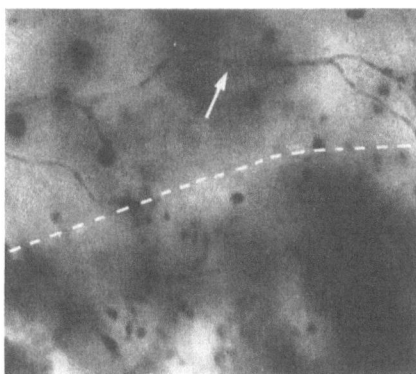

Fig 39. Surface preparation showing thin fibres (arrow) running outside the sensory epi-thelium of the crista. These fibres can be followed into the sensory epithelium, where they join the intraepithelial plexus. The border between the sensory epithelium and the surrounding epithelium is indicated by a broken line. $\times$ 940

The fibres in the intraepithelial plexus are not, however, strictly confined to the sensory epithelium. Thin fibres leave the plexus, to form arches outside the outer limits of the epithelium where they also branch (Fig. 39).

No differences in the innervation of the different cristae were observed. There were also no differences in the innervation of the utricular and canalicular sides of the sensory epithelium on the same crista.

2. Innervation of the Macula utriculi

The macula utriculi is innervated by myelinated fibres of different calibres. Thick fibres run to the striola region, penetrating the basement membrane more perpendicularly than those which supply the peripheral parts of the macula. The striola is further innervated by thin myelinated fibres, and medium-sized and thin fibres enter the sensory epithelium outside the striola. The sensory cells in a peripheral, marginal zone are innervated by fibres which lose their myelin sheath more centrally, and then run, either intra- or subepithelially, for a long distance in a peripheral direction as unmyelinated fibres. The number of myelinated fibres per unit surface area in the striola is larger than in any other area.

As in the crista, the nerve fibres lose their myelin sheath shortly before, or at the point of penetration of the basement membrane. Each of the thick fibres appears to innervate a small number of type I cells lying relatively close together in the striola. The medium-sized fibres can be followed over long distances, and often innervate fairly widely scattered type I cells. On the maculae, more fre-quently than on the cristae, several type I cells (2—5) are seen in a common chalice. This is most noticeable in the striola. All type I cells innervated by the same afferent nerve fibre showed the same morphological polarization.

Impregnation of the intraepithelial fibres was more difficult to obtain on the macula utriculi than in the sensory epithelium on the crista. It was therefore not possible to get satisfactory information about the pattern of innervation of type I and type II cells in the different regions.

On the macula utriculi there is also a well-developed intraepithelial plexus of very thin fibres. Compared with the plexus on the crista, this one seems to be located slightly higher in the epithelium, and to have more direct contact with the plasma membrane of type II cells and the chalices of type I cells. Otherwise the appearance is the same. It spreads over the whole macula utriculi, including the striola region. The thin fibres cross the striola, and one and the same fibre appears to make contact with sensory structures on both sides of the line of polarization.

Impregnation of the fibres was too irregular to allow of conclusions in regard to possible regional differences in the structure of the intraepithelial plexus. No relationship could be demonstrated between the intraepithelial spaces and the nerve fibres in the sensory epithelium.

In one specimen, very thin fibres were seen just outside and parallel to the lateral outer boundary of the macula utriculi. They had a slightly wavy course through this supposedly secretory epithelium (see p. 23). Branches passed into the intraepithelial plexus of the sensory epithelium, where the threads frequently divided, often with swellings at the point of division.

3. Innervation of the Macula sacculi

The macula sacculi is also innervated by fibres of different calibres. Regional differences in the distribution of these fibres are, however, less marked than on the macula utriculi.

The striola is innervated by thick fibres and a smaller number of thin fibres, and the density of thick fibres is greatest in the anterior part. The area inferior to the striola also receives thick fibres, but they are fewer than in the striola itself. The area superior to the striola is innervated exclusively by thin myelinated fibres.

On the macula sacculi, impregnation of the intraepithelial fibres was unsuccessful with the nerve stain used. Nothing can therefore be said about the pattern of distribution of these fibres.

D. Discussion

The innervation of the vestibular sensory epithelia is very complicated. In addition to afferent innervation through fibres which end as chalices around type I cells and through bud-shaped nerve endings on type II cells, it is assumed that the sensory epithelium has a somatic efferent innervation, and possibly also an autonomic innervation.

The modification of Maillet's (1963) method used in this investigation, gave fairly selective staining of the supposed efferent system in the organ of Corti (Engström et al., 1966b); and electron microscopy of the vestibular sensory epithelia showed that it also has a special affinity for the terminals of the efferent fibres. However, the staining was not selective, since there could also be impregnation of the nerve chalices of type I cells. In addition the technique generally gave less good staining of neural structures in the vestibular apparatus than in the cochlea. As with the different silver impregnation methods (Poljak, 1927b; Kolmer, 1927), this method gave incomplete impregnation of the intraepithelial nerve structures.

The present findings are in agreement with previous observations on the distribution of fibres of different calibre to central and peripheral regions of the sensory
epithelium on the crista (Cajal, 1908, 1909—1911; Lorente de Nó, 1926, 1931;
Poljak, 1927b; Wersäll, 1956); and they show that the thickest fibres end as
nerve chalices around a small number of type I cells on the central part of the
crista. Lorente de Nó (1926) found that all the cells innervated by one and the
same thick fibre in this region were orientated in one row, parallel to the semicircular duct, at right angle to the long axis of the crista; but this arrangement
of sensory cells has not been confirmed either by Poljak (1927b) or by the present
author.

The observations revealed that the number of myelinated nerve fibres per unit
surface area was considerably higher in the peripheral regions than in the central
region of the sensory epithelium on the crista, but since a systematic count of
these fibres was not made, a correlation with the number of sensory cells in these
regions cannot be given. As mentioned previously (p. 50), the density of sensory
cells was also considerably higher in the peripheral regions of the sensory epithelium on the crista than it was centrally. When determining the number of subepithelial fibres, however, it should be remembered that these may branch before
losing their myelin sheath, thus increasing their number peripherally. This is true,
for example, of the efferent myelinated fibres (Gacek, 1966). Even if a subepithelial branching of the myelinated fibres does not appear to occur extensively,
it should nevertheless be born in mind in making quantitative assessments of the
number of fibres in the various nerve branches to the vestibular sensory regions.

The thickest fibres on the macula utriculi run to the striola, an observation
in agreement with those of Lorente de Nó (1926, 1931), Poljak (1927b) and Engström and Wersäll (1958a). Each fibre innervates a small number of type I cells
which lie relatively close to one another, and the number of myelinated nerve
fibres per unit surface area is higher in the striola than in the peripheral areas.
Two ore more type I cells were often seen in a common nerve chalice, especially
in the striola. Cells in such chalices always had a uniform morphological polarization. Referring to this, it is interesting that in the lateral line organ in the frog,
cells orientated in one direction are innervated by certain nerve fibres, while cells
orientated in the opposite direction are innervated by others (Görner, 1963).
These observations thus appear to indicate that the same afferent fibre innervates
only cells with the same morphological polarization. Within those areas of the
vestibular sensory epithelium in which the majority of the cells are orientated
in a definite direction, a limited number of the cells may show deviations in their
morphological polarization; and a knowledge of the afferent innervation of these
latter cells would therefore be of great interest.

The myelinated fibres to the macula sacculi also differ in size, but opinions
vary regarding the regional differences in the distribution of these fibres. Lorente
de Nó (1926) and Hardy (1934) divided the macula sacculi into three regions, and
described a decreasing fibre calibre in the posterior direction. On the other hand,
Poljak (1927b) observed thick fibres also in the posterior parts of the macula
sacculi. This is in agreement with the present findings, which also showed that
even if thick fibres could be followed throughout the whole striola, the density
of such fibres was greatest in the anterior part. Likewise the area inferior to the

striola also receives thick fibres, but their density is less than that in, and immediately around, the striola. The observation that the area superior to the striola is innervated by thin fibres is, however, in agreement with the findings by Lorente de Nó (1926) and Hardy (1934). It was in this region that the thin longitudinally running fibres from Voit's anastomosis ended (p. 27).

Unfortunately our knowledge of the intraepithelial pattern of branching of the afferent fibres is still incomplete. This is especially true for the innervation of type II cells. The present investigation did not reveal whether the different afferent bud-shaped nerve endings making contact with a type II cell originated from several fibres or from only one. The fact that there is often a large number of nerve endings in contact with one type II cell, however, indicates that they originate from several fibres, and this would be in agreement with the conclusions of Wersäll (1956, 1960, 1966) and Spoendlin (1966 b).

The synaptic contact between the non-vesiculated, supposedly afferent nerve endings and the sensory cells has been made the subject of a number of recent electron microscopical studies (Engström et al., 1965; Spoendlin, 1965, 1966 b; Wersäll et al., 1966, 1967; Smith, 1967), and two types of synapses have been described between type I cells and their nerve chalices. In the area of contact there is a thickening of both plasma membranes, and synaptic bars and vesicles are present in the adjacent part of the sensory cell. Invaginations of the sensory cells are also observed, in which the distance between the plasma membrane of the sensory cell and that of the nerve chalice is greatly reduced. It is possible that this synaptic contact signifies an electrical transmission (Engström et al., 1965). Only synaptic structures indicating chemical transmission have been seen between type II cells and their non-vesiculated nerve endings.

A number of authors have previously described an intraepithelial plexus of nerve fibres on the cristae and maculae. The present findings on the crista showed that this plexus was best developed in the peripheral regions, and this is in agreement with the findings by Lorente de Nó (1926), Poljak (1927 b) and Wersäll (1956).

Lorente de Nó (1926) found no intraepithelial plexus on the macula utriculi in an area corresponding approximately to the striola, but Poljak (1927 b) and Engström and Wersäll (1958a), on the contrary, observed a plexus here also. The present observations confirm this finding, and show that these fibres can be followed over long distances, and that the same fibre appears to make contact with sensory structures on both sides of the polarization line in the striola. The nerve staining method used was, however, not successful in demonstrating a similar intraepithelial plexus on the macula sacculi.

Different views exist as regards the types of nerve fibres participating in the formation of this intraepithelial plexus. Lorente de Nó (1926) and Poljak (1927 b) described it as being composed both of collaterals from thicker fibres and of independent thin fibres, and Wersäll (1956) was of the opinion that it consisted of the medium-sized and thin fibres which innervated type II cells. With the nerve staining technique used in the present study, an intraepithelial plexus of very thin fibres was revealed, and these fibres were usually more heavily impregnated than the thick fibres. The possibility cannot be excluded that thin afferent fibres to type II cells, and even autonomic fibres may participate in the formation of this intraepithelial plexus. However, it is probable that the majority of the fibres in the plexus are of the somatic efferent type, since electron microscopy has shown that the bud-shaped swellings of the fibres in the plexus correspond to the vesiculated nerve endings which, according to Engström (1958), represent boutons.

There has, however, been considerable disagreement as to whether the vesiculated nerve endings are true terminals, or whether they are but regional swellings of the nerve fibres — boutons en passant. Iurato and Taidelli (1964) have recently shown that the majority of the richly vesiculated nerve endings are boutons en passant and they described boutons only 0.5—4 μ apart on the same fibre. Furthermore, each of the presynaptic fibres could have several synapses with the same nerve chalice, as well as with larger nerve fibres and type II cells. Iurato and Taidelli's description of the branching and innervation pattern of these supposedly efferent fibres is in agreement with the present observations based on light microscopy. They show that each fibre in the intraepithelial plexus has numerous swellings either at the place of division of the fibre or along its course, which obviously correspond to the boutons en passant. Such swellings were also seen at the end of a fibre here forming boutons terminaux.

Previous histological, histochemical and electrophysiological observations have provided evidence that there is an efferent innervation of the vestibular sensory regions. One indication of this is that some of the myelinated fibres in the peripheral branches of the vestibular nerve degenerate after a lesion of it (Petroff, 1955; Rasmussen and Gacek, 1958; Gacek, 1960, 1966). According to Gacek (1966), the number of these efferent fibres is about 400 in the cat. They are relatively thin, with a diameter of 2 to 5 μ.

Histochemical studies have further shown that acetylcholinesterase is present in the vestibular nerve and its branches and also intraepithelially. This is taken as an indication of an efferent innervation of the vestibular apparatus (for references see p. 72). The acetylcholinesterase activity disappeared or decreased peripherally after a lesion of the nerve (Ireland and Farkashidy, 1961; Gacek et al., 1965). The histochemical methods, however, have been too rough to show the intraepithelial branching pattern of the acetylcholinesterase-containing fibres, but certain regional differences in acetylcholinesterase activity have been pointed out. Thus, the activity on the crista was higher peripherally than centrally (Hilding and Wersäll, 1962; Nomura et al., 1965). On the macula sacculi less acetylcholinesterase is present in the striola than at the periphery (Nomura et al., 1965). Rossi and Cortesina (1965 b) showed that efferent fibres in the rabbit came from different parts of the medulla oblongata, including certain areas of the ipsilateral Deiters' nucleus.

Electrophysiological observations have shown that efferent impulses can be recorded from some of the vestibular nerve fibres (Schmidt, 1963), and stimulation with direct current of an area corresponding to the ipsilateral vestibular nuclei resulted in changes of action potentials in some fibres (Trincker, 1965). This efferent innervation appears to exert a central control on activity in the sensory organs, and most authors agree that the efferent system has an inhibitory action. However, it should be mentioned that Sala (1965) is of the opinion that the efferent fibres can show an excitatory as well as an inhibitory effect, depending on the functional state of the receptors.

Although all the above mentioned findings obviously show that there is an efferent innervation of the vestibular sensory cells, there is still lacking any histological proof that the fibres in the intraepithelial plexus, described in detail in the present study, are efferent. Thus the plexus remained intact after a lesion of the vestibular nerve (Gacek, 1960), and in his electron microscopical study Spoendlin (1966a) observed no changes in the vesiculated nerve endings in the vestibular sensory epithelia one year after destruction of the VIIIth nerve. Such terminals had, however, completely disappeared in the organ of Corti.

The present findings do not allow any final conclusions as to whether the vestibular apparatus is innervated by autonomic fibres. Andrzejewski (1955) and Spoendlin (1965) have previously demonstrated a perivascular network of adrenergic fibres along the labyrinthine artery. In addition Spoendlin (1965) and Spoendlin and Lichtensteiger (1966), using Falck and Hillarp's method, showed thin adrenergic threads which could be followed below the sensory epithelium on the crista and the macula utriculi. These authors claimed that the threads corresponded to the thin unmyelinated fibres, which have been demonstrated by electron microscopy both in the branches of the vestibular nerve and subepithelially between the myelinated fibres (Wersäll, 1956). Wersäll (1960) was not able to follow these fibres into the sensory epithelium, and suggested therefore that the fibres ended in the capillary layer. Spoendlin (1965), however, found no relationship between the supposed subepithelial adrenergic fibres and the blood vessels, but was of the opinion that some of the unmyelinated fibres penetrated the basement membrane and were located between the supporting cells. The present study is in agreement with the latter observations since it showed very thin unmyelinated fibres which often had no relationship to blood vessels or myelinated fibres. Some of them could be followed into the sensory epithelium where they joined the intraepithelial plexus. The possibility cannot be excluded that these fibres are autonomic.

With the nerve stain used, it was also possible to impregnate fibres running in the walls of the membranous labyrinth, just outside the sensory epithelium. These fibres lie parallel to the outer border of the sensory epithelium in the utricle and often passed in loops outside the sensory epithelium in the ampullae. They had branches which gave off collaterals to the intraepithelial plexus. The present author was not able to show where these fibres originated, and they were only seen in the immediate vicinity of the sensory epithelium. Similar fibres have been described by Andrzejewski (1955) and by Palumbi (1954), the latter author indicating that they could be of sensory or autonomic origin, and that they could make contact with secretory cells, connective tissue elements and blood vessels. He ventures the possibility that these fibres may have a regulating effect on the endolymphatic pressure and "determining reflexes of secretory and vasomotor order".

This discussion clearly shows that the innervation of the vestibular sensory regions is very complicated, and that there are still many problems unsolved. One of these is whether, in additon to the afferent and supposed efferent fibres, there are also autonomic fibres in the sensory epithelium. Likewise our knowledge of the branching of the afferent nerve of type I cells in the different regions is incomplete, and still less is known of the pattern of afferent innervation of type II cells. The possible regional differences in the structure of the efferent fibre system should also be studied in more detail, and future investigators should attempt to correlate the innervation of the sensory epithelium with the morphological polarization of the sensory cells in the corresponding area.

VIII. Form and Structure of the Statoconial Membranes and the Cupulae

A. Introduction

The sensory regions in the vestibular labyrinth are covered by a substance considered to be of a gelatinous nature. On the maculae this substance contains numerous crystals of high specific gravity, the statoconia. The gelatinous substance together with the statoconia is here called the *statoconial membrane*. This shows regional differences in structure to which little attention has been paid, even if differences have been described as regards the size of the individual crystals, the thickness of the layer of crystals (Lorente de Nó, 1926), and also the structure of the gelatinous substance in which the crystals are embedded (Werner, 1933).

The gelatinous substance covering the sensory epithelium on the cristae is called the cupula and does not contain crystals. It is much thicker than the statoconial membrane, and, according to Steinhausen (1933) it normally extends from the sensory epithelium right up to the roof of the ampulla and outwards to its side walls. Under the influence of fixatives it usually shrinks considerably and thus alters in both shape and size.

B. Material and Methods

Temporal bones from guinea pigs were used for these investigations. After fixation in 1.5 % osmium tetroxide or 10 % formalin, the membranous labyrinth was exposed as described on p. 9. Small squirts of fluid through a thin pipette were usually sufficient to remove the statoconial membrane in one piece from the sensory epithelium. It was transferred to a slide for detailed study under the light and phase-contrast microscope (see p. 9). The following methods were used for these studies.

1. Study of the shape of the statoconial membranes by light and phase-contrast microscopy, using low power magnification (Fig. 11a, c).

2. X-ray photography of the statoconial membranes (Fig. 40a, b).

3. Detailed study of the size and shape of the crystals. This was carried out after exerting pressure on the cover glass, thus splitting the statoconial membrane with resultant freeing of crystals from the gelatinous substance.

4. After exposure to air for about 10 minutes, for complete drying, the statoconial membrane was covered with Canada balsam. This made the crystal layer more transparent, and made it possible to study the size of the individual crystals and their position in the statoconial membrane (Fig. 42a).

5. A few drops of 2 % HNO_3 were added to whole statoconial membranes on a slide. This resulted in decalcification of the crystals so that the structure of the gelatinous substance surrounding the crystals could be seen (Fig. 42b).

The cupulae were placed in glycerin on a slide and studied directly under the phase-contrast microscope.

Some maculae and cristae with their statoconial membranes and cupulae respectively were embedded in Epon, sectioned and studied by light and phase-contrast microscopy.

C. Observations

After a wide opening has been made into the vestibule, the statoconial membranes are seen in their normal positions over the sensory epithelia (Figs. 2, 4). In direct light they are shining white. They have the same shape as the sensory epithelia, which they cover completely.

In non-fixed specimens the statoconial membranes are adherent to the sensory epithelia and the sensory hairs, and the gelatinous substance surrounding the crystals feels fairly viscous. After fixation the gelatinous substance attains a more solid state. Because of this, it is usually possible to isolate the statoconial membrane in one piece from the underlying sensory epithelium.

The basic structural features of the statoconial membranes over the macula utriculi and over the macula sacculi are the same. Closer study, however, reveals obvious regional differences. Such differences can be demonstrated in regard to the thickness of the crystal layer, the size of the crystals and the structure of the gelatinous substance, so justifying separate descriptions of the statoconial membrane on the macula utriculi and on the macula sacculi.

1. Statoconial Membrane of the Macula utriculi

Radiology shows that the thickness of the crystal layer on the macula utriculi is greatest laterally and slightly less posterior and anterior to the striola (Figs. 40b, 44). In the striola the crystal layer is very thin. Medial to the striola the crystal layer is generally of medium thickness, but this varies somewhat in the different regions. In the striola, and to a lesser extent outside this area as well, there are small crater-like depressions of the crystal layer where this is especially thin (Figs. 40b, 45a).

The crystals are located in the upper part of the gelatinous substance and they lie in several layers, very close together (Fig. 45a). The crystals have an oblong, hexagonal form (cp. Fig. 41), and the length varies from $1/2$ to 30 μ. Most of the crystals are from 4 to 8 μ long. Sometimes a "nucleus" is seen (cp. Fig. 41).

Constant regional differences are observed in regard to the size of the crystals (Figs. 44a, 45a). The upper layer of the statoconial membrane contains small crystals. Similar small crystals are also seen in a narrow marginal zone above the outer border of the sensory epithelium. Further basally, i.e. nearer to the sensory hairs, the crystals vary in size. The crystals in the pars externa (p. 44), outside the striola, are generally very large (Figs. 44a, 45a). The largest crystals, about 30 μ long, are seen lateral to the striola where the crystal layer as a whole is thickest. The long axis of the crystals in this area often lies perpendicular to the surface of the sensory epithelium. In the striola the crystals are very small, with a length of about 1 μ. Medial to the striola they are medium in size. In a marginal, medial area, corresponding to that part of the macula utriculi in which the sensory

Fig. 40a and b. Microradiographs of the statoconial membranes of the macula sacculi and the macula utriculi in the guinea pig. The crystal layer in the striola of the macula sacculi (S in a) is thicker than that on either side. The crystal layer in the striola of the macula utriculi (S in b) is thinner than that on either side. Note "craters" (arrows) mainly in the striola of the statoconial membrane of the macula utriculi. a and b $\times$ 95

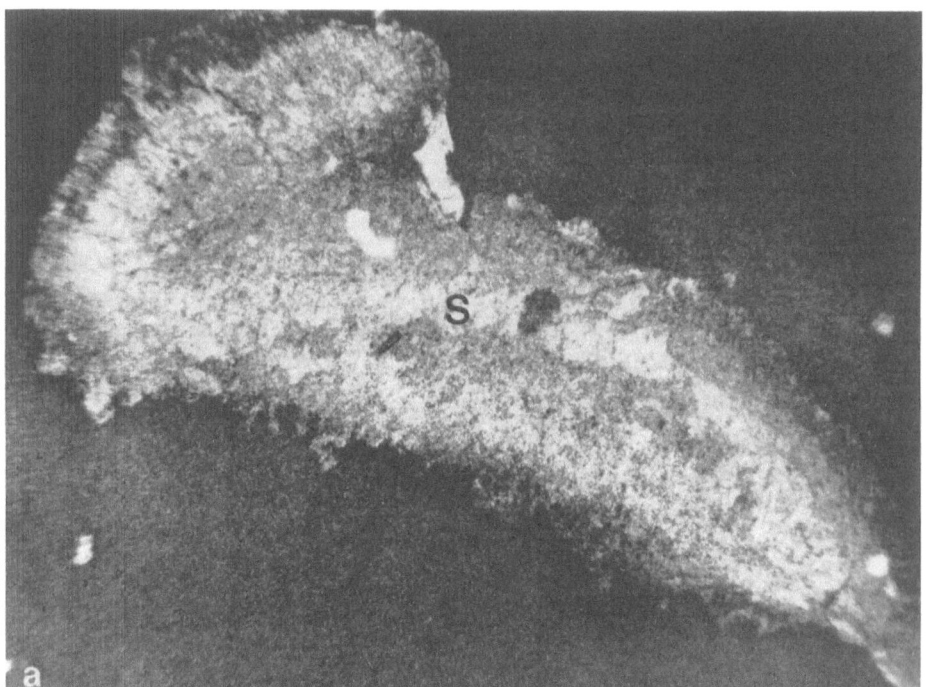

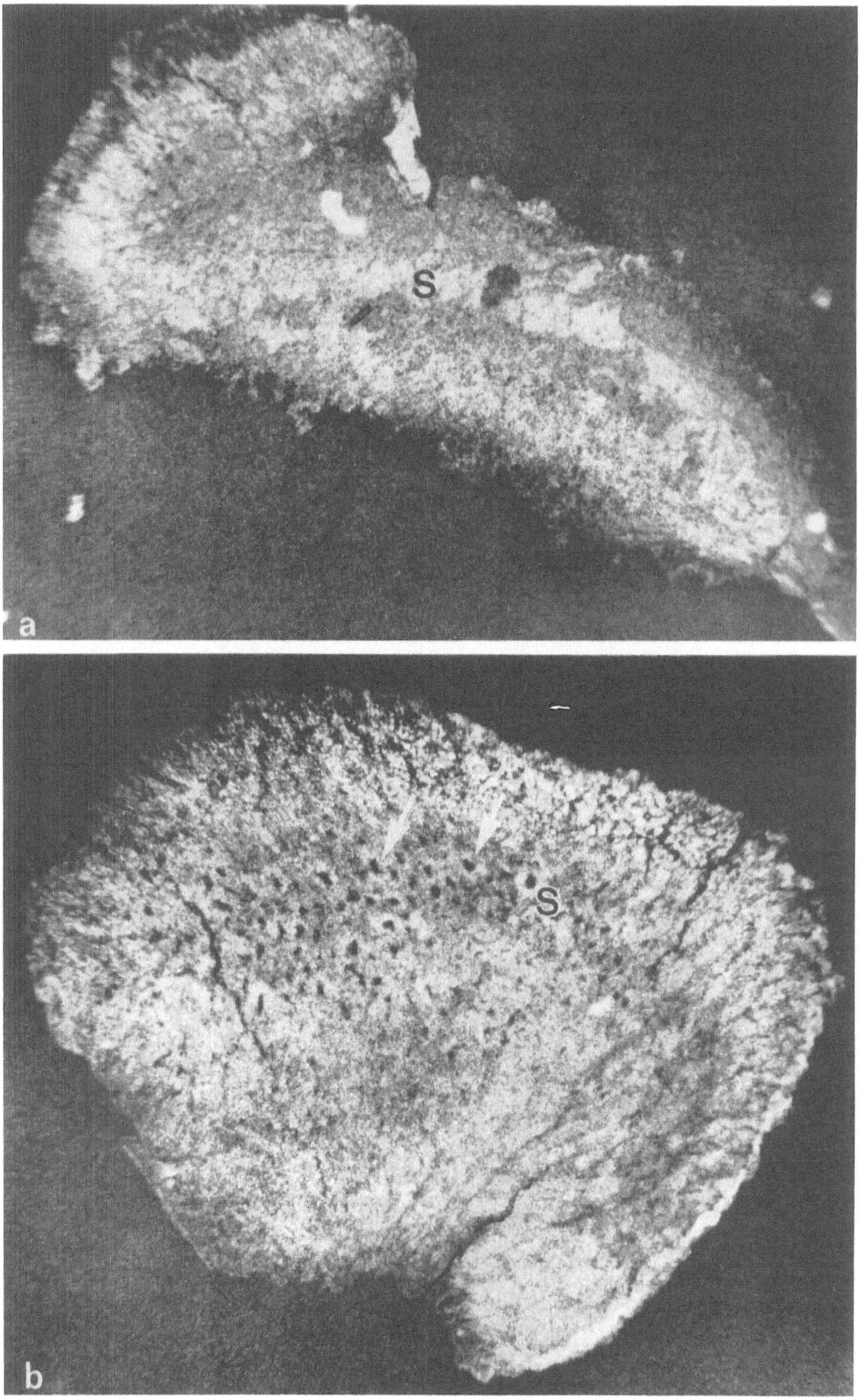

Fig. 40 a and b

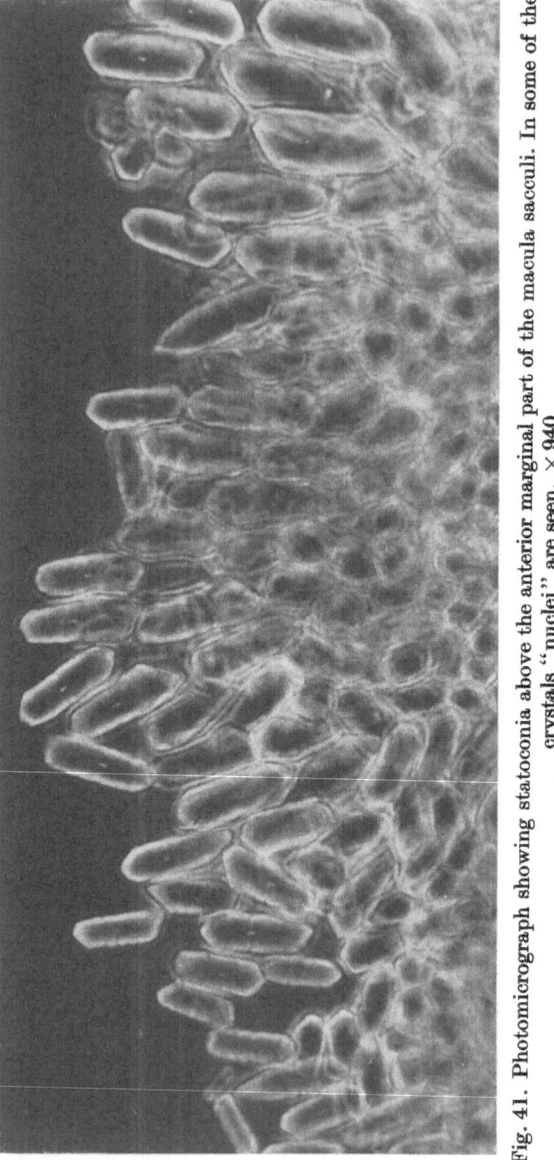

Fig. 41. Photomicrograph showing statoconia above the anterior marginal part of the macula sacculi. In some of the crystals "nuclei" are seen. × 940

cells are scattered (see p. 43), the crystal layer is extremely thin and the crystals very small (Figs. 43, 44a).

The upper surface of the crystal layer has a characteristic contour in some areas, seen especially well with side illumination (Fig. 43). This is due to small grooves and elevations on the surface of the crystal layer. The long axis of these contours coincides with the morphological polarization of the sensory epithelium in the area in question. This orientation is most pronounced over the pars interna of the macula utriculi, and is not seen in the striola.

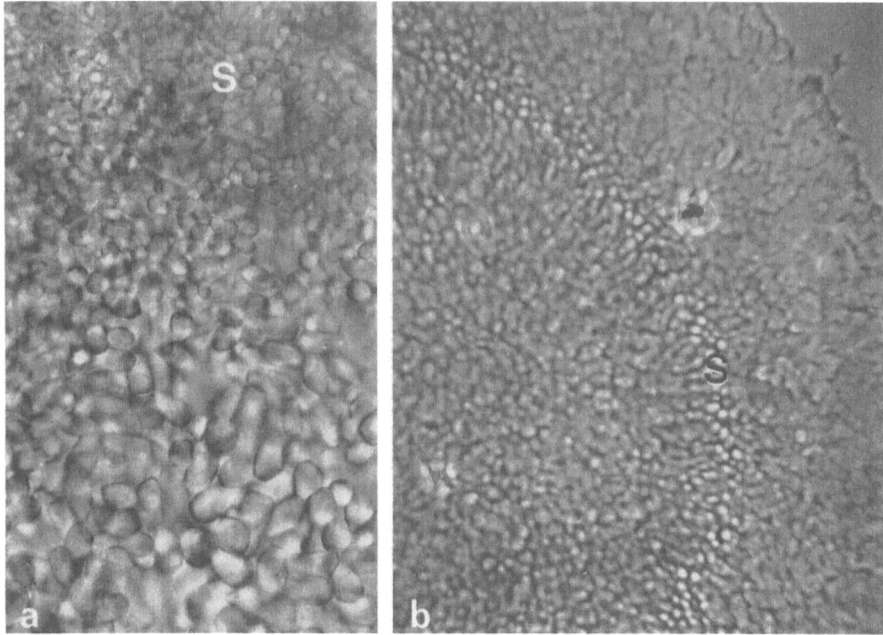

Fig. 42a and b. In a the lower surface of the statoconial membrane of the macula sacculi in the guinea pig is seen. Small crystals are found in the striola (s). b Surface preparation showing the statoconial membrane of the macula utriculi, as seen after decalcification. In the striola (s), a reinforcement of the reticular structure is observed. a × 780, b × 155

Studies of isolated statoconial membranes gave the impression that there was a groove in the lower surface of the crystal layer, opposite the surface of the sensory epithelium in the striola region. This lower layer appears to be further away from the surface of the sensory epithelium here than does the crystal layer on the two sides of the striola.

In fixed specimens there are also regional differences in the structure of the gelatinous substance. This was especially studied in decalcified statoconial membranes. Under the phase-contrast microscope, the gelatinous substance is revealed as a fibrillar structure forming a network. This network is much stronger and coarser in the striola (cp. Fig. 42b, s) than it is peripherally, where the substance appears more structureless. In non-decalcified specimens the network is more strongly impregnated with the Giemsa stain in the striola than at the periphery.

In some cases statoconial membranes with an unusual appearance were seen. They contained a smaller number of giant crystals of very variable shape. It was especially easy to study the gelatinous substance in areas where there were no crystals. Its appearance was then the same as that after decalcification. By separating the sensory epithelium together with the gelatinous substance, it was also possible to study the relationship of the sensory hairs to the gelatinous substance. In the striola there were broad clearings in the gelatinous substance, and these formed canal-like spaces for the hair bundles of the sensory cells. Peripherally these clearings were not seen so distinctly, and they also had a much smaller extension than in the striola.

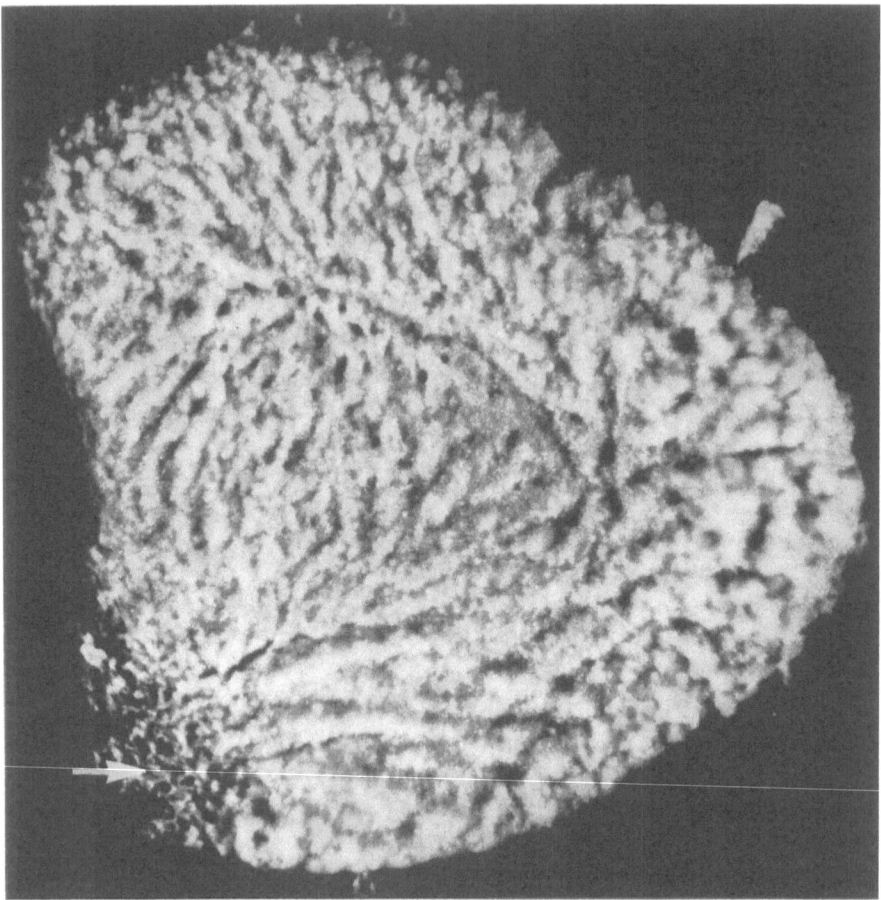

Fig. 43. Surface view of the upper side of the statoconial membrane of the macula utriculi in the guinea pig. The crystal layer shows furrows and ridges, the long axis of which generally corresponds to the morphological polarization in the actual area. Note the extremely thin crystal layer above the medial part of the macula utriculi (arrow). × 114

2. Statoconial Membrane of the Macula sacculi

There are also regional differences in the structure of the statoconial membrane on the macula sacculi. In an area corresponding to the striola, there is a snowdrift-like elevation of the upper surface of the crystal layer (Figs. 3b, 11c, 45b). In this area the crystal layer is thicker than on the two sides of the striola (Figs. 40a, 45b).

The crystals are arranged in several layers. Their shape is the same as on the macula utriculi and the length varies from $^1/_2$ to 15—20 μ, most of the crystals being from 3 to 7 μ long. On the macula sacculi there are also constant regional differences in the position of large and small crystals (Figs. 42a, 44b, 45b). The upper layer of the statoconial membrane is composed of small crystals. Similar crystals also form a narrow marginal zone at the outer borders of the sensory

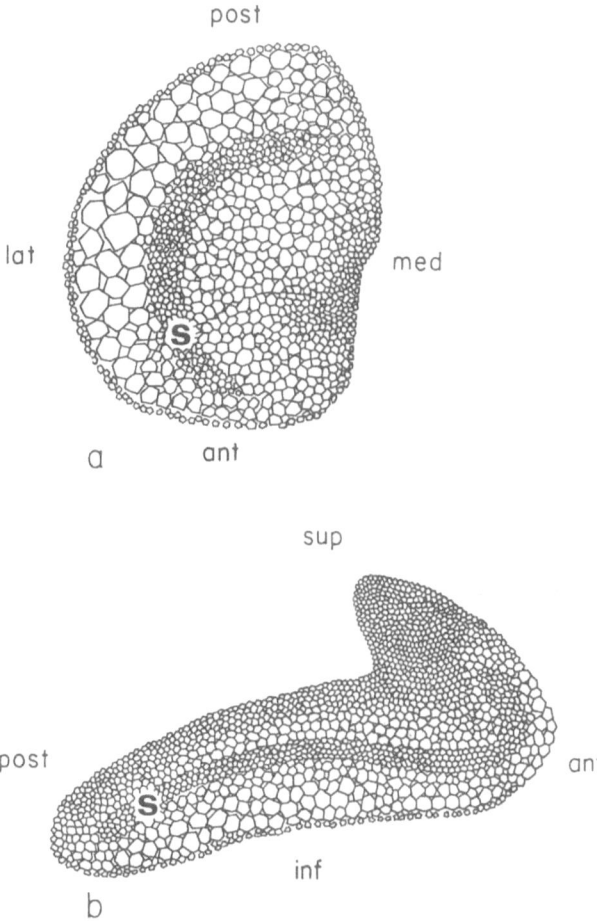

post

lat

med

S

a   ant

sup

post

S

ant

inf

b

Fig. 44. Diagram illustrating regional differences in the size of the crystals. The length of the statoconia in the guinea pig varies from $^1/_2\,\mu$ to about 30 μ. The largest crystals are found in the areas of the statoconial membranes covering the lateral part of the macula utriculi and the antero-inferior part of the macula sacculi. In the striola (S) of both maculae the crystals are small

epithelium, except in the anterior part of the macula sacculi. In this region there are relatively large crystals most marginally (Fig. 41).

The largest crystals are seen inferior to the striola, especially in the posterior part of the pars externa. In an area corresponding to the snowdrift-like thickening — the striola — all the crystals are very small (Figs. 42a, 45b). In a superior direction from the striola they first increase slightly in size, and then gradually decrease.

After fixation, the structure of the gelatinous substance on the macula sacculi is the same as on the macula utriculi. There is a similar reinforcement of the gelatinous substance in the striola compared with the peripheral regions.

Fig. 45a, b illustrates the regional differences mentioned in the structure of the statoconial membranes on the macula utriculi and on the macula sacculi.

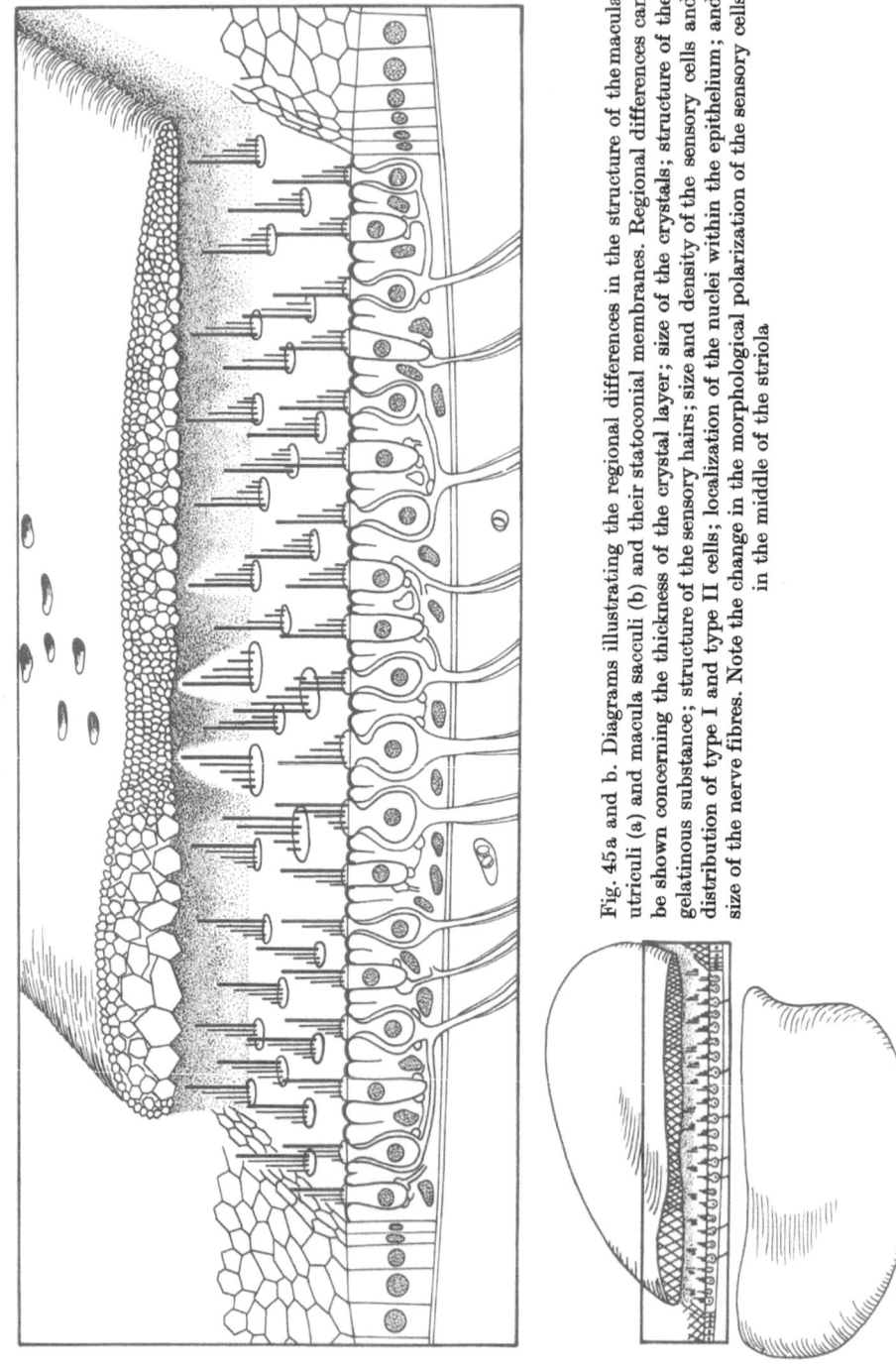

Fig. 45a and b. Diagrams illustrating the regional differences in the structure of the macula utriculi (a) and macula sacculi (b) and their statoconial membranes. Regional differences can be shown concerning the thickness of the crystal layer; size of the crystals; structure of the gelatinous substance; structure of the sensory hairs; size and density of the sensory cells and distribution of type I and type II cells; localization of the nuclei within the epithelium; and size of the nerve fibres. Note the change in the morphological polarization of the sensory cells in the middle of the striola

3. Cupulae

With careful preparation, the cupula is seen in its normal position in the ampulla, riding on the crista. Even if the cupula shrinks during fixation in

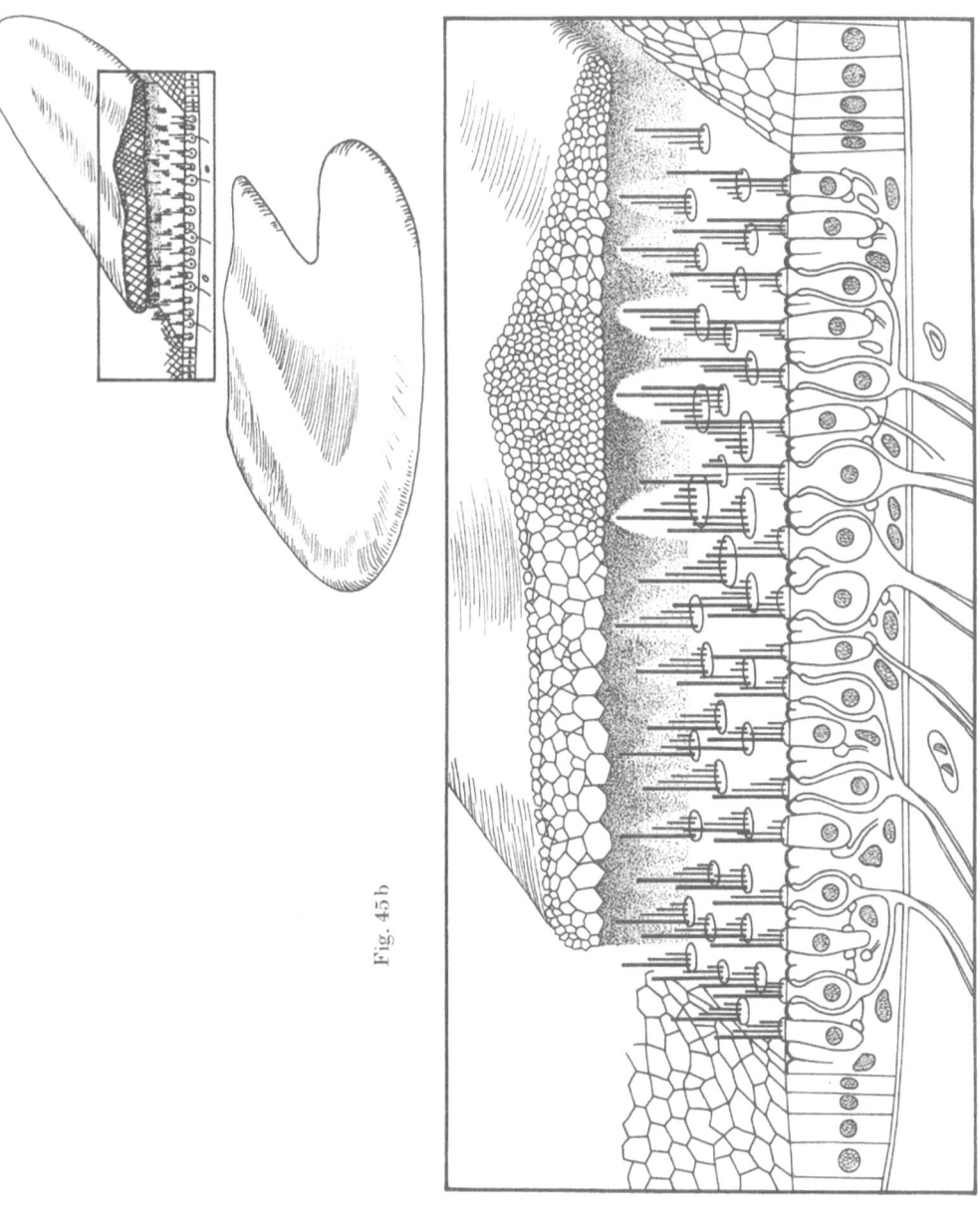

Fig. 45 b

osmium tetroxide, it keeps its shape well. A concave surface covers the sensory epithelium on the crista and two surfaces face the utricular and canalicular sides of the ampulla respectively (Fig. 8). Two borders face towards the sidewalls with the plana semilunata and one surface faces the roof of the ampulla. Phase-contrast microscopy shows parallel fibrillar structures in the long axis of the cupula. These structures run from the upper part of the cupula, bordering the roof of the ampulla, and almost down to the basal part which borders the sensory epithelium. Most basally, the substance of the cupula becomes increasingly structureless and difficult to observe. There is an obvious difference between, on the one hand, the fairly diffuse border between the cupula and the surface of the sensory epithelium, and, on the other hand, the distinct border between the cupula and the other regions mentioned above.

D. Discussion

Compared with the abundant literature on the structure and innervation of the sensory epithelia, it is astonishing that little interest has been paid to the statoconial membranes and cupulae. A review of the large number of names used to describe the supraepithelial structures on the maculae appears, however, to give a different impression[2]. It is obvious that many of these names have been connected both with the crystal membrane as a whole, and with its components. Furthermore, the same name has also been used to identify different structures, which adds a complication. The terms statoconial membrane and statoconium are used in the present study and these also agree most closely with the nomenclature in *Nomina anatomica* (1966) — membrana statoconiorum and statoconium respectively.

Kolmer (1927) was of the opinion that the statoconia in mammals consisted of aragonite. More recent investigations have shown that this crystal occurs in amphibia (Brandenberger and Schintz, 1945; Carlström *et al.*, 1953), but that in man as well as in various mammals, birds and sharks, the crystals are composed of calcite (Brandenberger and Schintz, 1945; Carlström *et al.*, 1953; Carlström and Engström, 1955; Sasaki and Miyata, 1955). Calcite and aragonite are two forms of calcium carbonate, but show differences in physical properties such as specific gravity, refractive index, hardness and crystalline form (Carlström and Engström, 1955)[3].

Calcite crystals have a high specific gravity (about 2.71). They are hexagonal, with slightly arched sides and pointed ends, and electron microscopy of statoconia in man (Carlström and Engström, 1955) shows that the outline of the pointed ends is always uneven. According to Werner (1940), they are covered by a membrane of organic substance.

Several authors have described a "nucleus" in the crystals, and Kolmer (1927) was of the opinion that this formed the starting point for crystal formation. Crystals with a "nucleus" were also seen occasionally in the present investigation. Its appearance varied to some extent, and this agrees with the description of Engström (1951). Carlström *et al.* (1953), however, did not succeed in studying these nuclei by electron microscopy of decalcified statoconia.

Reports vary on the length of the crystals. Igarashi (1966) described 7—8 μ long crystals in the squirrel-monkey, but Carlström *et al.* (1953) found that the length of the statoconia in man varied between 1 and 20 μ. This corresponds

2. Otolith, otolith membrane, otolithic membrane, otoconium membrane, statolith, statolith membrane, statoconium membrane, statoconial membrane and statoconium otolith are the different terms used by various students for the crystal membrane as a whole. The gelatinous substance has been described as the otolith membrane, otolithic membrane, statolith membrane, statoconium membrane, and macularium, and the individual crystals have been called otolith, otoconium, statolith, statoconium, ossiculith and microstatolith.

3. The statoconia in cyclostomosis contain calcium phosphate (Carlström and Engström, 1955).

closely with the present findings in the guinea pig, where the length of the crystal lies between $1/2$ and 30 μ. The observations have shown, furthermore, that statoconia are best studied by light/phase contrast microscopy after isolation of the statoconial membrane. Such studies reveal that the length of the statoconia shows greater variations than appear from perusal of the relevant literature.

The statoconia lie close together in several layers, and not in one layer as often reported. Lorente de Nó (1926) described regional differences in the thickness of the crystal layer, and in the distribution of large and small crystals on the macula utriculi of the mouse. His division of the statoconial membrane corresponded to his division of the sensory epithelium into different regions with relation to its innervation (see p. 72). Laterally on the macula utriculi Lorente de Nó found a very thick crystal layer and large crystals, in agreement with Werner's (1933) and the present author's observations. Werner (1933) found, however, that most of the rest of the statoconial membrane contained a medium thick layer of small crystals, while Lorente de Nó and the present author found here a considerably more differentiated structure. In the striola (see p. 52) small crystals form a thin layer. Medially on the macula utriculi Lorente de Nó (1926) observed a thick layer of small crystals. The present study, however, shows that the thickness of the crystal layer is less here than it is lateral to the striola, and that the crystals generally are of a medium size. In a marginal medial area, over the part of the macula utriculi where the sensory cells are very scattered (see p. 43), there is, however, a very thin layer of small crystals. As mentioned previously, it was also observed that both the upper layer of the statoconial membrane and a narrow completely marginal zone consisted of small crystals.

Neither Lorente de Nó nor Werner described regional differences in the thickness of the crystal layer or in the size of the crystals on the macula sacculi. Ades and Engström (1965), however, examined different animal species and found that the crystals were accumulated in a snowdrift-like thickening along a curved central line on the macula sacculi. This corresponded to the dividing line for the morphological polarization of the sensory epithelium. They found a similar central thickening of the statoconial membrane on the macula utriculi, a finding which was supported by Iurato (1967). The present observations in the guinea pig confirm Ades and Engström's observations on the macula sacculi but not on the macula utriculi, since X-ray studies showed a thinning of the crystal layer corresponding to the striola of that macula.

The present study also revealed regional differences in the size of the crystals on the macula sacculi in the guinea pig. In the snowdrift-like thickening of the crystal layer — which corresponds to the striola — the crystals were very small. The largest crystals were found inferior to the striola, and they were smaller than the largest crystals on the macula utriculi.

Furthermore, the investigation showed that the lower surface of the crystal layer in the striola of the macula utriculi was more remote from the sensory epithelium than the crystal layer on each side of the striola. The illustrations by Lorente de Nó (1926, p. 99, Fig. 21, mouse) and by Werner (1933, p. 705, Fig. 12, hen), support this observation. On the macula sacculi, on the other hand, the lower surface of the crystal layer could lie nearer to the sensory epithelium than the crystal layer on each side of the striola.

The gelatinous component of the statoconial membrane is considered by some authors to have the same histochemical characteristics as the cupula and tectorial membrane (Wislocki and Ladman, 1955; and others), but according to Iurato and de Petris (1967) its composition is not completely known. Several authors agree, however, that the gelatinous component of the statoconial membrane and the cupula is formed by the supporting cells of the sensory epithelium (Retzius, 1884; Kolmer, 1927; Wersäll, 1956; Flock, 1964; Smith, 1967; and others).

Werner (1933) described regional differences in the structure of the gelatinous component of the statoconial membrane in the guinea pig and rabbit. There was a special development of this substance in an area corresponding to a central, curved line above the maculae. This area, which sometimes could be stained with haematoxylin, was the region for Werner's first definition of the striola (see p. 52). The present findings support Werner's observations and show that the gelatinous substance has a net-like structure, especially well developed in the striola. The structure also forms canal-like spaces for the sensory hairs, a finding in agreement with that of Kolmer (1927); and the canals are broader and more conspicuous in the striola than peripherally.

The gelatinous substance was chiefly studied in decalcified statoconial membranes, but had the same structure in the statoconial membranes where the crystals appeared to be dissolved and had been replaced by a smaller number of giant crystals. Some of these giant crystals were hexagonal and had a normal appearance. It is possible that Lindeman's (1967) observation of crystals up to 50 μ was based on measurement of these giant crystals, which do not generally seem to be present. The largest crystals, however, had a very irregular shape and bore little resemblance to normal statoconia. It is most probable that this finding was the result of artefacts, and that there had been a post mortem dissolution of the crystal structure with subsequent re-crystallization. The changes in the statoconial membrane thus seemed to be most pronounced in specimens which, after fixation in osmium tetroxide, had been stored for some days in alcohol in a refrigerator. Referring to this, it is interesting that Werner (1940) described dissolution of the crystals when specimens were stored in a cold place, and formation of giant crystals at higher temperatures. He concluded, however, that previous fixation in acetic acid was the primary cause of the changes.

Without previous fixation the cupula is completely transparent and cannot be studied directly. By injecting stains into the membranous labyrinth of living fish, however, Steinhausen (1933) was able to demonstrate the form of the cupula under normal conditions, and the way in which its position in the ampulla was influenced by artificial currents in the semicircular canals. He was of the opinion that, under normal conditions, the cupula extended from the surface of the sensory epithelium to the roof of the ampulla and touched the lateral walls of the ampulla with the plana semilunata parietally. It thus closed the lumen of the ampulla completely to the endolymph, an assumption which agrees with that of Trincker (1965). However, this view of the normal form and size of the cupula is not shared by others. Thus, the illustrations by Kolmer (1924), Wersäll (1956), and Wersäll and Flock (1965) show that the cupula neither reaches to the roof of the ampulla, nor to its lateral walls. The form and size of the cupula, as found in the present study, agreed with the illustrations of the latter authors, but it should be stressed that these studies were made on fixed specimens. The rather angular shape of the cupula found after fixation does not, however, exclude the possibility that, under normal conditions, it follows the more regular rounded shape of the ampullar walls, and Werner (1932) actually showed that both volume and shape of the cupula were very easily influenced by, for example, different fixatives. On the other hand, he believed that under normal conditions the cupula covered the cross section of the ampulla completely. Although decisive conclusions cannot be reached, it is important to realize that the cupula shrinks considerably under the influence of most fixatives, and that changes in the shape and size of the cupula, often assumed to be due to pathological conditions, are probably a result of the fixation.

Electron microscopy has shown that the cupula is composed of fine filaments (Wersäll, 1956), and according to Iurato and de Petris (1967) these filaments are 30—40 Å thick. They are not arranged in any definite direction, and "the fine cottony appearance is that of a dispersed reticular gel" (loc. cit., p. 217). In addition, there are, throughout the whole length of the cupula, fine canals (Kolmer, 1927; Werner, 1940) in which the hair bundles of the sensory cells are located. According to Wersäll (1956), the canals in the basal part of the cupula have a diameter of 3—5 μ, those in the upper part 1 μ or less.

In the present investigation, an attempt was made to elucidate the relationship of the cupula to the surface of the sensory epithelium. The longitudinal, parallel, coarsely fibrillar structures which were seen in phase-contrast microscopy running through the cupula, terminated at some distance above the surface of the sensory epithelium, and the gelatinous substance then gradually became more structureless and diffuse. No sharp demarcation could be seen basally, and it was not possible to determine with certainty whether or not the gelatinous substance made contact with the epithelial surface. Kolmer (1926) assumed that the gelatinous substance extended down to the supporting cells, but it was so flimsy basally that the connection with the supporting cells was usually broken as a result of shrinkage during fixation. The present studies on the cupulae in the two vertical semicircular ducts in the cat are interesting in this connection. The sensory epithelium of the cristae of these semicircular ducts was divided in two by a bar of indifferent epithelium (see p. 21). The gelatinous substance appeared to be absent above this area, and the impression was thus obtained that the cupula was divided in two, with a free space between. However, a closer study of this middle zone showed therein the presence of gelatinous substance which was of the same structureless appearance as that immediately above the surface of the sensory epithelium, and there did not seem to be an absolute division between the two halves of the cupula.

On the macula the gelatinous substance is most fully developed just under the crystal layer, and its relationship to the surface of the sensory epithelium is, as on the cristae, difficult to determine. Kolmer (1927) meant that the gelatinous substance made contact with the epithelial surface. However, a sub-cupular space has often been described (Wersäll, 1956; Igarashi, 1966; and others) as a slit between, on the one hand the surface of the sensory epithelium and, on the other hand, the cupula or the lower surface of the gelatinous substance of the stato-conial membrane; but opinions vary concerning the depth of this sub-cupular space. This is obviously due to different degrees of shrinkage during fixation, since Kolmer (1926) has shown that the gelatinous substance immediately over the surface of the sensory epithelium is especially susceptible to fixatives. The sub-cupular space often seen in sections is therefore probably largely an artefact due to the fixation and embedding, and this assumption is supported by the findings made in the present investigation.

IX. General Discussion

In order to understand the function of the vestibular apparatus and also to assess whether or not there are pathological changes in it, it is essential to have accurate knowledge of its normal anatomy. Our knowledge of the vestibular sensory regions is, however, still incomplete. This is probably due mainly to the fact that these sensory regions are enclosed in a hard shell of bone, which makes a closer examination difficult.

In the present century most of the morphological studies of the vestibular apparatus have been carried out on sections of decalcified and embedded temporal bones. These studies have indeed increased our knowledge of the structure of the vestibular sensory regions, but such investigations also have their clear limi-tations, as has recently been emphasized by Fernandez (1958a, b), Igarashi et al. (1967) and others. Some of these deficiencies are also discussed briefly in chapter I. They are often the cause of difficulties in assessing whether or not pathological changes are present in the vestibular sensory regions. We have a very limited knowledge of the changes in the structure of these regions in different types of vertigo which are assumed to be due to changes in the receptor organ, and we know nothing about the possible changes therein with increasing age. The diffi-culties in assessing with this technique the loss of a few sensory cells in the vestib-ular sensory epithelia have also been pointed out more recently (see for example Smith, 1965).

During the past decade, electron microscopical studies have shed new light on the structure of the vestibular sensory regions. These studies have shown that the structure of the vestibular sensory epithelia is far more complicated than was previously realized. It should, however, be pointed out that, used alone, the electron microscope is of limited value. Indeed, a detailed knowledge of the struc-ture of the various parts of the sensory regions at the light microscopical level is essential before a systematic study of the different areas is made with the electron microscope. The present investigation is an example of this.

The microdissection technique, which was used especially by Retzius (1881a, 1884) with such mastery, has recently been shown to be very useful for structural studies of the organ of Corti under normal and pathological conditions. In the present study an attempt has been made to show that a similar technique should also provide a useful supplement to the study of sections of embedded material from the vestibular sensory regions. The technique is discussed in detail in chapter II. An especially important aspect of the technique is that it permits identification of virtually all the sensory and supporting cells within a given sensory region. It therefore seems well suited to quantitative determinations of cells in both normal and pathological sensory epithelium. The method is reliable, orientation of the specimen is possible, and each cell can be related to other cells or structures. These criteria are essential for a systematic study of corresponding regions in different animals. It should, however, be stressed that, as with light/phase-contrast microscopy of sections or electron microscopy used alone, this method cannot satisfy every requirement. A combination of the different methods will therefore always give a more accurate, detailed picture of the structure of the sensory regions under normal or pathological conditions.

In the present investigation the distinct regional differences in the structure of the vestibular sensory regions are especially clearly shown. Regional differences were revealed both in the structure of the statoconial membranes and in the structure and innervation of the sensory epithelia. These findings have been discussed in their respective chapters, where the findings are related to those of other investigators.

A characteristic pattern of the striola of the maculae and of the central regions of the cristae was that the majority of their sensory cells had a large free surface, and that the density of these cells was less than in the peripheral regions. The concentration of type I cells was higher in the striola than in the periphery, a finding in agreement with that of Engström and Wersäll (1958a) and Spoendlin (1965). Wersäll (1956) and Spoendlin (1965) concluded that type I cells on the cristae were located chiefly in the central regions, while type II cells dominated the peripheral regions (for details see p. 54). The present observations, however, do not support these findings, since approximately the same percentage of type I and II cells was found in the central and peripheral regions of the crista. The present observations, however, confirmed Werner's (1933) findings in the striola of the guinea pig, regarding the position of the nuclei of the sensory and supporting cells in two distinctly different layers. A further agreement with Werner's (1933) description was the finding of intraepithelial spaces in the striola of the macula utriculi. Similar spaces were also seen, though to a lesser extent, in the striola of the macula sacculi. A concentration of large osmiophilic granules, probably lipid droplets, was observed in the supporting cells in both the striola and the central regions of the crista in the guinea pig.

As discussed in chapter VII there were also distinct regional differences in the innervation of the vestibular sensory regions. The thickest fibres went to the central areas of the crista and to the striola of the maculae, most noticeably on the macula utriculi.

Light and phase-contrast microscopy also demonstrated regional differences in the appearances of the sensory hairs. Peripherally in the sensory regions the hairs were thinner and more graceful, and on the cristae they were also longer, than they were centrally. In the striola and the central areas of the cristae, especially on the cells with a large free surface, the hairs were thicker and more club-shaped. These regions should be studied in more detail by electron micro-scopy.

Study of surface specimens showed that the hair bundles on the sensory cells in the striola were located in wide canal-like spaces in the gelatinous substance of the statoconial membrane. This gelatinous substance had a distinctly reticular appearance, much coarser in the striola than peripherally. In the guinea pig the crystal layer in the striola on the macula sacculi was thicker, and that in the striola on the macula utriculi was thinner, than on each side of the striola.

Furthermore, distinct regional differences in the size of the crystals were seen in the guinea pig. The crystals were small in the striola of both maculae.

The structure of the vestibular sensory regions strongly indicates that the striola and the central regions of the cristae differ functionally from the peripheral areas. However, electrophysiological studies have not yet demonstrated any such functional differentiation. As pointed out by Brodal (1966), the anatomical data exceed what has so far been clarified by physiological studies.

Some authors have attempted to attribute a functional significance to the morphological findings. Lorente de Nó (1926) was of the opinion that the innervation of the central regions was finer and more differentiated than that of the peripheral regions; and Poljak (1927a) compared the innervation of the central and peripheral areas of the vestibular sensory regions with the innervation of the fovea centralis and the peripheral regions of the retina.

The present observations clearly show that the consistency of the statoconial membrane changes during fixation, and that it is easy to remove the membrane from the sensory epithelium after treatment with various agents. All studies of fixed statoconial membranes give the impression that this is an entity which, under the influence of linear acceleration, can be displaced as a single, fairly firm structure across the macula. Unfixed specimens, however, show that their consistency is entirely different from that of fixed membranes. Such preparations show that it is difficult to remove the membrane from the sensory epithelium, and that it is very viscous. Therefore, it is easier to conceive that the regional differences in the structure of the statoconial membrane signify a functional differentiation, and it is probable that the different areas of the statoconial membrane in vivo exert a selective effect on the underlying sensory cells.

Opposite conditions were found not only regarding the thickness of the crystal layer in the striola, but also regarding the patterns of polarization in the macula utriculi and the macula sacculi. Each macula was divided into two areas by an imaginary line in the middle of the striola, with different polarization on each side of this line. The sensory cells were morphologically polarized towards this line in the macula utriculi, whilst in the macula sacculi the situation was reversed. In the present study the two halves of each macula were called the pars interna and the pars externa. This subdivision of the maculae seems justified, the more so since the morphological polarization obviously must be of great functional importance. This differentiated structure should be borne in mind in future studies on the structure and function of the maculae.

In the present study, especial emphasis was placed on showing the structural differences between the striola and the peripheral regions of the maculae. However, the investigation also revealed other details in the structure of the maculae. Medially on the macula utriculi, where the outer border of the sensory epithelium folds in, only scattered sensory cells are present. The statoconial membrane over this area in the guinea pig consists of a thin crystal layer, thinner than over any other region, and the crystals are very small. In addition, the sensory cells in the pars interna of the macula utriculi are morphologically polarized according to a pattern in which the directions of polarization diverge fan-like in a posterior, lateral and slightly anterior direction from this marginal, medial area.

It was interesting to note that the orientation of the cells in a certain region could be seen not only in the arrangement of the sensory hairs, but also in the cellular pattern and the structure of the statoconial membrane. Thus, the free surface of the supporting cells was oval, the upper surface of the statoconial membrane showed grooves and elevations, and the long axis of the pattern was in the same direction as the morphological polarization of the sensory cells in the area.

In the mammals examined, the basic features of the structure of the maculae were the same. In this laboratory, however, Rosenhall (1968) showed regional differences in the structure of the macula utriculi in birds which differed from those in mammals, a finding in agreement with Werner's (1938) description. The structure of the macula utriculi in birds thus seems to be even more differentiated than that in mammals. However, even if the findings indicate that the morphological conditions are simpler in mammals, there are many structural details which are not easy to understand.

Considering the fine structural differentiation of the vestibular sensory regions, it seemed to be of interest to study the effect of noxious agents on the vestibular sensory epithelia. This type of investigation should show whether the sensory epithelium could be considered as a uniform one — from the point of view of vulnerability — or whether there are also regional differences in sensitivity. In the guinea pig, streptomycin applied to the middle ear and parenteral injections of kanamycin produced the same patterns of degeneration in the vestibular sensory regions (Lindeman, 1967, 1969). In the first place, the sensory epithelium of the cristae was more vulnerable than that of the macula utriculi, and this again was more vulnerable than that of the macula sacculi. Furthermore, distinct regional differences in vulnerability were demonstrated within each individual sensory region. Thus the sensory cells in the central areas of the crista were more vulnerable than those in the periphery of the crista. On the maculae the sensory cells on the striola were more vulnerable than those at the periphery. No explanation can yet be given as to why the large cells in the striola and the central areas of the cristae, chiefly type I cells, were most vulnerable. However, according to Spoendlin (1966a), there seems to be some correlation between the intensity of protein metabolism in the different cells and their sensitivity to streptomycin. It seems not unreasonable to assume that the regional differences in the vulnerability of the cells also reflect differences in the function of the cells. Systematic studies with the electron microscope are certainly advisable, partly to attempt to identify the point of attack of these ototoxic antibiotics in the sensory cells, and partly to try to find, on an ultrastructural level, a possible morphological correlation with the regional differences in vulnerability mentioned above.

Studies of the effect of ototoxic antibiotics on the vestibular sensory epithelia showed that degeneration of sensory cells in the mildest cases was confined to the central parts of the cristae and the striola of the maculae. It seems important to stress this. In future experimental studies, or when assessing the sensory epithelium from patients with vestibular disturbances of assumed peripheral origin, it should be realized that possible degeneration of sensory cells may be confined to small areas, for example the central regions of the crista or the striola of the maculae. Thus, sections made in the long axis of the striola may give incorrect information about damage to the macula as a whole, and sections made outside the striola may fail to show possible degenerative changes. However, application of different noxious agents does not always seem to produce the same pattern of degenerative changes in the vestibular sensory regions. Thus Winther (1968), after X-ray irradiation of guinea pigs, demonstrated a pattern of degeneration which differed from that found after treatment with ototoxic antibiotics.

It is of interest to know whether the structural differentiation of the vestibular sensory regions is also reflected in the pattern of the central projection of the nerve fibres. Apparently the only investigation relevant to this problem is that of Lorente de Nó (1926). He followed fibres from one area of the macula utriculi, which corresponded essentially to the striola, to large cells in the vestibular ganglion, whilst the rest of the macula utriculi was found to be projected on to small cells (see chapter IV). If these findings are correct, one should expect a similarly differentiated projection from the central and peripheral regions of the macula sacculi and cristae. This remains to be examined in future studies.

The demonstration of the pattern of polarization on the maculae also raises other questions. The dividing line for polarization runs along the middle of the striola. Does this mean that the two halves of the striola project themselves on to different regions centrally? Is there any correlation between the polarization pattern and the projection of the sensory epithelia on to the vestibular ganglion and the vestibular nuclei? Referring to the differentiation of the sensory cells into two types, it can also be asked whether there is a functional difference between type I and type II cells, and also whether there is a difference in their central projection. Before these questions can be answered, a more detailed examination is needed of the intraepithelial course of the nerve fibres. The branching pattern of the fibres has not been revealed in the present study, and little as yet is known about it. It is, however, complicated. An example of this is that the same afferent fibre can innervate both a type I and a type II cell.

X. Summary

The structure of the vestibular apparatus has been studied on temporal bones of the guinea pig, rabbit, cat, squirrel monkey and man. This study has been based on microdissection and on sections made for light/phase-contrast and electron microscopy. Microdissection gives a three dimensional view of the membranous labyrinth, and the form and relationships, of the ducts and sacs appear clearly. A technique for the preparation of the vestibular sensory regions is described. After exposure of the maculae and cristae, the statoconial membranes and cupulae are removed for closer study. The entire sensory epithelia are then peeled off and mounted in glycerin or Canada balsam. Light and phase-contrast microscopy of these surface preparations permit identification of every sensory and supporting cell within a given sensory region. Blood vessels and nerves within the sub-epithelial tissues can be similarly studied. The method is fast and reliable and allows of exact orientation, which is a prerequisite for systematic studies of corresponding areas in different animals.

Clear regional differences are revealed in the structure of the sensory epithelia and their innervation, and in the structure of the statoconial membranes. In the macula sacculi and the macula utriculi, a centrally located area, the striola, is structurally differentiated from the remaining regions. In the macula sacculi of the guinea pig, this area constitutes about 13 per cent of the sensory epithelium, in the macula utriculi about 8 per cent. In the striola, most of the sensory cells have an especially large free surface, and the density of the sensory cells (number of cells per surface area) is less than it is peripherally. Furthermore, the striola in the guinea pig is characterized by a particular arrangement of the nuclei of the sensory and supporting cells, by a concentration of osmiophilic granules in the supporting cells, and by intraepithelial spaces.

Quantitative observations on five preparations of each macula in the guinea pig reveal the following average dꝼ ta: Macula sacculi: surface area 0.495 mm², number of sensory cells 7,560. Macula utriculi: surface area 0.541 mm², number of sensory cells 9,260. The average density of sensory cells in the macula utricuil is slightly higher than in the macula sacculi. Observations in different guinea pigs show that, in the striola, about two-thirds of the sensory cells are of type I and one-third of type II, whereas these cells have an almost equal distribution at the periphery. The sensory epithelia of the cristae can be divided structurally into central and peripheral regions. In the central regions most of the sensory cells are characterized by a large free surface, and the density of the sensory cells is distinctly less here than it is peripherally. Furthermore, in the central regions a concentration of large osmiophilic granules is noticed in the supporting cells. In the guinea pig, no apparent regional differences in the distribution of type I and type II cells are seen; type I cells constitute about 60 per cent of the number of sensory cells, both centrally and peripherally. Observations on three preparations from each crista reveal the following average data: Crista anterior: surface area 0.369 mm², number of sensory cells 5,442. Crista posterior: surface area 0.337 mm², number of sensory cells 5,430. Crista lateralis: surface area 0.383 mm², number of sensory cells 5,688.

Each sensory cell is morphologically polarized, with regard to the location of the kinocilium in its relationship to the stereocilia. On the crista lateralis all sensory cells are morphologically polarized towards the utricle, on the cristae of the two vertical semicircular ducts away from the utricle. Each macula is divided into two areas by an arbitrary, curved line, situated in the middle of the striola, generally with opposite morphological polarization on each side of this line. The areas are called the pars interna and the pars externa. On the macula utriculi the sensory cells are morphologically polarized towards the dividing line, whereas on the macula sacculi they are polarized away from it. On either end of the striola, irregularities in the pattern of polarization are seen. In the macula sacculi of the rabbit and cat, a narrow irregular zone, located antero-inferiorly, shows a deviation from the general pattern in that the sensory cells are polarized towards the above-mentioned dividing line.

In some regions, there is seen a distinct orientation in the cellular pattern of the sensory epithelium and in the structure of the statoconial membrane. This is in accordance with the direction of polarization in the actual area.

In the pars interna of the macula utriculi, the morphological polarization of the sensory cells follows a characteristic pattern. From a marginal, medially located area, the directions of polarization spread fan-like posteriorly, laterally and somewhat anteriorly. In this medially-situated area, the sensory cells lie scattered amongst the supporting cells, and the crystal layer of the overlying statoconial membrane is extremely thin.

Differences in structure of the sensory hairs are found in the different sensory regions. On the cristae, the hairs in the peripheral parts of the epithelium are longer than those in the central part. The stereocilia in the central areas of the cristae and in the striola of the maculae, are thicker and more club-shaped than the stereocilia at the periphery.

The course and ramifications of the vestibular nerve have been studied, and special attention has been paid to the innervation of the macula sacculi. This macula is innervated by myelinated fibres from both the saccular and Voit's nerve. In all species an overlapping of fibres from the two branches is seen. Furthermore, in the guinea pig bundles of thin myelinated fibres can be followed from Voit's nerve far into the main part of the macula sacculi.

The vestibular epithelia are innervated by myelinated nerve fibres of different calibre. The fibres which run to the central parts of the cristae and to the striola of the maculae, are thicker than those which run to the periphery. In the sensory epithelium a plexus of very fine fibres is seen. These fibres are provided with beads, which probably represent boutons. Each of the thin fibres appears to make contact with a large number of type II cells, with the nerve chalices of type I cells and with thicker intraepithelial fibres. Below the sensory epithelium, unmyelinated fibres are observed, some of which can be followed into the sensory epithelium. Moreover, thin unmyelinated fibres are seen in the walls of the membranous labyrinth, just outside the sensory epithelia of the cristae and the macula utriculi.

The statoconial membranes show regional differences in the size of the crystals, thickness of the crystal layer, and structure of the gelatinous substance enveloping the crystals. In the guinea pig, the crystal layer in the striola of the macula utriculi is thinner than that on either side, whereas on the macula sacculi, the crystal layer in the striola is thicker than that on either side. The size of the crystals varies from about $1/2$ μ to about 30 μ in the guinea pig. The largest crystals are found above the lateral part of the macula utriculi and the antero-inferior part of the macula sacculi. In the striola of both maculae, only small crystals are seen. The gelatinous substance shows a fibrillar appearance, much coarser in the striola than at the periphery.

The present study reveals a highly differentiated structural organization of the vestibular sensory regions. These observations indicate that the different areas are functionally dissimilar, a suggestion which is further supported by the findings of clear regional differences in the sensitivity of the vestibular sensory cells to ototoxic antibiotica.

Acknowledgement. The present investigation was carried out at the Ear, Nose and Throat Department, Sahlgrenska Hospital, University of Gothenburg.

I wish to express my gratitude to the Head of the Department, Professor Gösta Herberts, for his interest in the study and for providing good working facilities. Professor Hans Engström introduced me to the sensory regions of the inner ear. I am greatly indebted to him for his inspiring interest in the work and for critical advice. Throughout the study Dr. Göran Bredberg has offered me great help, and I am most greatful to him for stimulating discussions and advice.

My sincere thanks are due to Professor Fred Walberg and Professor Alf Brodal at the Anatomical Institute, University of Oslo, for their constructive criticism and generous help, particularly in the preparation of the manuscript.

I wish to express my gratitude to Dr. John Ballantyne for correcting the English text and for valuable advice. The skilful technical assistance and advice of Res. Engineer Anton Andersson, is greatfully acknowledged. For stimulating discussions I want to thank Dr. Bernhard Kellerhals, Dr. Roy Naessen, and my father, Associate Professor Henrik R. Lindeman.

Furthermore, I am indebted to Associate Professor Hans Röckert for assistance in taking the X-rays; to Dr. Susan Schanche for translation of the main part of the manuscript into English; and to Mrs. Nanti Arnborg for expert help in making the schematic drawings.

This study was supported in part by the Norwegian Research Council for Science and the Humanities; in part by the Physiological Psychology Branch, Office of Naval Research, Washington DC, under Contracts N 62558 4264 and F 61052 67 C 0090 with H. Engström. This support is gratefully acknowledged.

References

Ades, H. W., and H. Engström: Form and innervation of the vestibular epithelia. In: The role of the vestibular organs in the exploration of space, p. 23—41. NASA SP-77. National Aeronautics and Space Administration, Washington D.C., 1965.

Adrian, E. D.: Discharges from vestibular receptors in the cat. J. Physiol. (Lond.) 101, 389—407 (1943).

Alexander, G.: Zur Anatomie des Ganglion vestibulare der Säugethiere. S.-B. Akad. Wiss. Wien, Abt. I 108, 449—469 (1899).

— Zur Anatomie des Ganglion vestibulare der Säugethiere. Arch. Ohr.-, Nas.-, u. Kehlk.-Heilk. 51—52, 109—125 (1901a).

— Das Labyrinthpigment des Menschen und der höheren Säugethiere. Arch. mikr. Anat. 58, 134—181 (1901b).

— Makroskopische Anatomie der nervösen Anteile des Gehörorganes. In: Handbuch der Neurologie des Ohres, Bd. I, S. 1—100 (eds. G. Alexander und O. Marburg). Berlin u. Wien: Urban & Schwarzenberg 1924.

Andersen, H. C.: Passage of trypan blue into the endolymphatic system of the labyrinth. Acta oto-laryng. (Stockh.) 36, 273—283 (1948).

Andrzejewski, C.: Histologische Studien zur vegetativen und cerebralen Innervation des Innenohrs und seiner Gefäße beim Menschen und beim Hund. Z. Zellforsch. 42, 1—18 (1955).

— Weitere histologische Beobachtungen über die vegetative Innervation des membranösen Labyrinthes. Z. Zellforsch. 44, 427—440 (1956).

Anson, B. J., D. G. Harper, and T. R. Winch: The vestibular system: Anatomic considerations. Arch. Otolaryng. 85, 497—516 (1967).

Bairati, A.: Récentes connaissances sur la structure submicroscopique des organes du vestibule. Acta otolaryng. (Stockh.), Suppl. 163, 9—25 (1961).

—, and S. Iurato: The ultrastructural organization of "plana semilunata". Exp. Cell Res. 20, 77—83 (1960).

Ballantyne, J., and H. Engström: Morphology of the vestibular ganglion cells. J. Laryng. 83, 19—42 (1969).

Bast, T. H., and B. J. Anson: The temporal bone and the ear. Springfield (Ill.): Thomas 1949.

Békésy, G. v.: Shearing microphonics produced by vibrations near the inner and outer hair cells. J. acoust. Soc. Amer. 25, 786—790 (1953).

Benjamins, C. E.: A new peripherical network in the vestibular organ of the pike (Esox lucius). Acta oto-laryng. (Stockh.) 8, 60—64 (1925).

Bielschowsky, M., u. G. Brühl: Über die nervösen Endorgane im häutigen Labyrinth der Säugetiere. Arch. mikr. Anat. 71, 22—57 (1908).

Borghesan, E.: The reticular zone of the planum semilunatum. Acta oto-laryng. (Stockh.) 54, 27—32 (1962).

Brandenberger, E., and H. R. Schinz: Über die Natur der Verkalkungen bei Mensch und Tier und das Verhalten der anorganischen Knochensubstanz im Falle der hauptsächlichen menschlichen Knochenkrankheiten. Helv. med. Acta, Suppl. 16, 1—63 (1945).

Bredberg, G.: Cellular pattern and nerve supply of the human organ of Corti. Acta oto-laryng. (Stockh.), Suppl. 236, 1—135 (1968).

— H. Engström, and H. W. Ades: Cellular pattern and nerve supply of the human organ of Corti. A preliminary report. Arch. Otolaryng. 82, 462—469 (1965).

Brodal, A.: Anatomical aspects on functional organization of the vestibular nuclei. In: The role of the vestibular organs in space exploration, p. 119—139. NASA SP-115. National Aeronautics and Space Administration, Washington, D.C., 1966.

— O. Pompeiano, and F. Walberg: The vestibular nuclei and their connections, anatomy and functional correlations. The Henderson trust Lectures. Edinburgh and London: Oliver & Boyd 1962.

Burlet, H. M. de: Der perilymphatische Raum des Meerschweinchenohres. Anat. Anz. 53, 302—315 (1920).

— Zur Innervation der macula sacculi bei Säugetieren. Anat. Anz. 58, 26—32 (1924).

—, u. J. H. de Haas: Die Stellung der Maculae acusticae im Meerschweinchenschädel. Z. Anat. Entwickl.-Gesch. 68, 177—197 (1923).

— — Die Stellung der Maculae acusticae im Macasusschädel. Z. Anat. Entwickl.-Gesch. 71, 233—239 (1924).

—, u. J. M. Hoffmann: Beitrag zur Kenntnis der Maculae acusticae bei Säugetieren. Arch. Ohr.-, Nas.- u. Kehlk.-Heilk. 120, 233—255 (1929).

Cajal, S. R.: Neue Darstellung vom histologischen Bau des Centralnervensystems. Arch. Anat. Entwickl.-Gesch. 319—428 (1893).

— Asociación del método del nitrato de plata con el embrionario. Trav. Lab. Rech. biol. Univ. Madr. 3, 65—96 (1904).

— Terminación periférica del nervio acustico de las aves. Trav. Lab. Rech. biol. Univ. Madr. 6, 161—176 (1908).

— Histologie du système nerveux de l'homme et des vertébrés. Tome I. Paris: Maloine 1909—1911.

Carlström, D.: A crystallographic study of the vertebrate otoliths. Biol. Bull. 125, 441—463 (1963).

—, and H. Engström: The ultrastructure of statoconia. Acta oto-laryng. (Stockh.) 45, 14—18 (1955).

— —, and S. Hjorth: Electron microscopic and X-ray diffraction studies of statoconia. Laryngoscope (St. Louis) 63, 1052—1057 (1953).

Dahl, H. A.: Fine structure of cilia in rat cerebral cortex. Z. Zellforsch. 60, 369—386 (1963).

Dijkgraaf, S.: The functioning and the significance of lateral-line organs. Biol. Rev. 38, 51—105 (1963).

Dohlman, G. F.: Histochemical studies of vestibular mechanisms. In: Neural mechanisms of the auditory and vestibular systems, p. 258—275 (eds. G. L. Rasmussen and W. F. Windle). Springfield (Ill.): Thomas 1960.

— The mechanism of secretion and absorption of endolymph in the vestibular apparatus. Acta oto-laryng. (Stockh.) 59, 275—288 (1965).

— J. Farkashidy, and F. Salonna: Centrifugal nerve-fibres to the sensory epithelium of the vestibular labyrinth. J. Laryng. 72, 984—991 (1958).

—, and F. C. Ormerod: The secretion and absorption of endolymph. Acta oto-laryng. (Stockh.) 51, 435—438 (1960).

Engström, H.: Microscopic anatomy of the inner ear. Acta oto-laryng. (Stockh.) 40, 5—22 (1951).

— On the double innervation of the sensory epithelia of the inner ear. Acta oto-laryng. (Stockh.) 49, 109—118 (1958).

— The cortilymph, the third lymph of the inner ear. Acta neerl. Morph. 3, 195—204 (1960).

— The innervation of the vestibular sensory cells. Acta oto-laryng. (Stockh.), Suppl. 163, 30—40 (1961).

— Elektronenoptische Histologie des Innenohres. In: Hals-Nasen-Ohrenheilkunde, S. 148—166 (eds. J. Berendes, R. Link und F. Zöllner). Stuttgart: Thieme 1965.

Engström, H., H. W. Ades, and J. E. Hawkins Jr.: Structure and functions of the sensory hairs of the inner ear. J. acoust. Soc. Amer. **34**, 1356—1363 (1962).
— — — The vestibular sensory cells and their innervation. In: Modern trends in neuro-morphology. Symp. Biol. Hung., vol. 5, p. 21—41 (ed. J. Szentágothai). Budapest: Akadémiai Kiadó 1965a.
— H. W. Ades, and J. E. Hawkins Jr.: Cellular pattern, nerve structures and fluid spaces of the organ of Corti. In: Contributions to sensory physiology, p. 1—37 (ed. W. D. Neff). New York and London: Academic Press 1965b.
— —, and A. Andersson: Structural pattern of the organ of Corti. A systematic mapping of sensory cells and neural elements. Stockholm: Almqvist & Wiksell 1966b.
—, and S. Hjorth: On the distribution and localization of injected dyes in the labyrinth of the guinea pig. Acta oto-laryng. (Stockh.), Suppl. **95**, 149—158 (1950).
—, H. H. Lindeman, and H. W. Ades: Anatomical features of the auricular sensory organs. In: The role of the vestibular organs in space exploration, p. 33—44. NASA SP-115. National Aeronautics and Space Administration, Washington, D.C. 1966a.
—, u. B. Rexed: Über die Kaliberverhältnisse der Nervenfasern im N. stato-acusticus des Menschen. Z. mikr.-anat. Forsch. **47**, 448—455 (1940).
—, and J. Wersäll: Structure and innervation of the inner ear sensory epithelia. Int. Rev. Cytol. **7**, 535—585 (1958a).
— — The ultrastructural organization of the organ of Corti and of the vestibular sensory epithelia. Exp. Cell Res. **5**, 460—492 (1958b).
Ernyei, I.: Die Elemente der Nerven und Ganglien des inneren Ohres. Arch. Ohr.-, Nas.- u. Kehlk.-Heilk. **141**, 343—348 (1935).
Fawcett, D. W.: Cilia and flagella. In: The cell, vol. II, p. 217—297 (eds. J. Brachet and A. E. Mirsky). New York and London: Academic Press 1961.
— The cell. Its organelles and inclusions. Philadelphia and London: Saunders Comp. 1966.
Fernández, C.: Postmortem changes in the vestibular and cochlear receptors (guinea pig). Arch. Otolaryng. **68**, 460—487 (1958a).
— Postmortem changes and artefacts in human temporal bones. Laryngoscope (St. Louis) **68**, 1586—1615 (1958b).
Flock, Å.: Structure of the macula utriculi with special reference to directional interplay of sensory responses as revealed by morphological polarization. J. Cell Biol. **22**, 413—431 (1964).
— Electron microscopic and electrophysiological studies on the lateral line canal organ. Acta oto-laryng. (Stockh.), Suppl. **199**, 1—90 (1965).
— Transducing mechanism in the lateral line canal organ receptors. Cold Spr. Harb. Symp. quant. Biol. **30**, 133—145 (1966).
—, and A. J. Duvall: The ultrastructure of the kinocilium of the sensory cells in the inner ear and lateral line organs. J. Cell Biol. **25**, 1—8 (1965).
— R. Kimura, P.-G. Lundquist, and J. Wersäll: Morphological basis of directional sensitivity of the outer hair cells in the organ of Corti. J. acoust. Soc. Amer. **34**, 1351—1355(1962).
—, and J. Wersäll: A study of the orientation of the sensory hairs of the receptor cells in the lateral line organ of fish, with special reference to the function of the receptors. J. Cell Biol. **15**, 19—27 (1962).
— — Morphological polarization and orientation of the hair cells in the labyrinth and the lateral line organ. J. Ultrastruct. Res. 8, 193—194 (1963).
Gacek, R. R.: Efferent component of the vestibular nerve. In: Neural mechanisms of the auditory and vestibular systems, p. 276—284 (eds. G. L. Rasmussen and W. F. Windle). Springfield (Ill.): Thomas 1960.
— The macula neglecta in the feline species. J. comp. Neurol. **116**, 317—323 (1961).
— The vestibular efferent pathway. In: The vestibular system and its diseases, p. 99—116. (ed. R. J. Wolfson). Philadelphia: Univ. Pennsylvania Press 1966.
— Y. Nomura, and K. Balogh: Acetylcholinesterase activity in the efferent fibres of the stato-acoustic nerve. Acta oto-laryng. (Stockh.) **59**, 541—553 (1965).
—, and G. L. Rasmussen: Fiber analysis of the stato-acoustic nerve of guinea pig, cat, and monkey. Anat. Rec. **139**, 455—463 (1961).
Gernandt, B.: Response of mammalian vestibular neurons to horizontal rotation and caloric stimulation. J. Neurophysiol. **12**, 173—184 (1949).

Gibbons, I. R.: The relationship between the fine structure and direction of beat in gill cilia of a lamellibranch mollusc. J. biophys. biochem. Cytol. 11, 179—205 (1961).

Görner, P.: Untersuchungen zur Morphologie und Elektrophysiologie des Seitenlinienorgans vom Krallenfrosch (Xenopus laevis Daudin). Z. vergl. Physiol. 47, 316—338 (1963).

Gray, A. A.: The labyrinth of animals. Including mammals, birds, reptiles and amphibians. London: Churchill 1907.

Guild, S. R.: Observations upon the structure and normal contents of the ductus and saccus endolymphaticus in the guinea pig (Cavia cobaya). Amer. J. Anat. 39, 1—56 (1927a).

— The circulation of the endolymph. Amer. J. Anat. 39, 57—81 (1927b).

Hardy, M.: Observations on the innervation of the macula sacculi in man. Anat. Rec. 59, 403—418 (1934).

Hasse, C.: Der Bogenapparat der Vögel. Z. wiss. Zool. 17, 598—645 (1867a).

— Nachtrag zur Arbeit: Der Bogenapparat der Vögel. Z. wiss. Zool. 17, 646—654 (1867b).

Held, H.: Untersuchungen über den feineren Bau des Ohrlabyrinthes der Wirbeltiere. I. Zur Kenntnis des Cortischen Organs und der übrigen Sinnesapparate des Labyrinthes bei Säugetieren. Abh. math.-phys. Kl. sächs. Akad. Wiss. (Lpz.) 28, 1—74 (1902).

— Untersuchungen über den feineren Bau des Ohrlabyrinthes der Wirbeltiere. II. Zur Entwicklungsgeschichte des Cortischen Organs und der Macula acustica bei Säugetieren und Vögeln. Abh. math-phys. Kl. sächs. Akad. Wiss. (Lpz) 31, 193—294 (1909).

— Die Cochlea der Säuger und der Vögel, ihre Entwicklung und ihr Bau. In: Handbuch der normalen und pathologischen Physiologie, Bd. 11, S. 467—534. Berlin: Springer 1926.

Hilding, D., and J. Wersäll: Cholinesterase and its relations to the nerve endings the inner ear. Acta oto-laryng. (Stockh.) 55, 205—217 (1962).

Igarashi, M.: Comparative histological study of the reinforced area of the saccular membrane in mammals. BuMed Project MR 005.13-6001 Subtask 1, Report No. 101 and NASA Order No. R-93. Pensacola, Fla.: Naval School of Aviation Medicine, p. 1—14 (1964).

— Dimensional study of the vestibular end organ apparatus. In: The role of the vestibular organs in space exploration, pp. 47—53. NASA SP-115. National Aeronautics and Space Administration, Washington, D.C., 1966.

— B. R. Alford, and F. R. Guilford: The sectioning planes for study of temporal bones. Ann. Otol. (St. Louis) 76, 330—345 (1967).

Ireland, P. E., and J. Farkashidy: Studies on the efferent innervation of the vestibular end organs. Ann. Otol. (St. Louis) 70, 490—503 (1961).

Ishii, T., Y. Murakami, and K. Balogh: Acetylcholinesterase activity in the efferent nerve fibers of the human inner ear. Ann. Otol. (St. Louis) 76, 69—82 (1967).

Iurato, S.: The sensory cells of the membranous labyrinth. Arch. Otolaryng. 75, 312—328 (1962).

— Submicroscopic structure of the inner ear. Oxford and London: Pergamon Press 1967.

—, and S. de Petris: Otolithic membranes and cupulae. In: Submicroscopic structure of the inner ear, p. 210—218 (ed. S. Iurato). Oxford and London: Pergamon Press 1967.

—, and G. Taidelli: Relationships and structure of the so-called "much granulated nerve endings" in the crista ampullaris (as studied by means of serial sections). Electron Microscopy 1964. Proc. Third Europ. Reg. Conf. (Prague), vol. B, p. 325—326 (1964).

Johnsson, L.-G., and J. E. Hawkins Jr.: A direct approach to cochlear anatomy and pathology in man. Arch. Otolaryng. 85, 599—613 (1967).

Kaiser, O.: Das Epithel der Cristae und Maculae acusticae. Arch. Ohrenheilk. 32, 181—194 (1891).

Kimura, R., P.-G. Lundquist, and J. Wersäll: Secretory epithelial linings in the ampullae of the guinea pig labyrinth. Acta oto-laryng. (Stockh.) 57, 517—530 (1963).

—, and H. B. Perlman: Extensive venous obstruction of the labyrinth. B. Vestibular changes. Ann. Otol. (St. Louis) 65, 620—638 (1956).

—, and H. F. Schuknecht: Membranous hydrops in the inner ear of the guinea pig after obliteration of the endolymphatic sac. Pract. oto-rhino-laryng. (Basel) 27, 343—354 (1965).

Kohonen, A.: Effect of some ototoxic drugs upon the pattern and innervation of cochlear sensory cells in the guinea pig. Acta oto-laryng. (Stockh.), Suppl. 208. 1—70 (1965).

Kolmer, W.: Über die Endigungsweise des Nervus octavus. Zbl. Physiol. 18, 620—623 (1904).
— Mikroskopische Anatomie des nervösen Apparates des Ohres. In: Handbuch der Neurologie des Ohres, Bd. I, S. 101—174 (eds. G. Alexander und O. Marburg). Berlin u. Wien: Urban & Schwarzenberg 1924.
— Über das Verhalten der Deckmembranen zum Sinnesepithel der Labyrinthendstellen. Arch. Ohr.-, Nas.- u. Kehlk.-Heilk. 116, 10—26 (1926).
— Gehörorgan. In: Handbuch der Mikroskopischen Anatomie des Menschen, vol. III, p. 250—505 (ed. W. v. Möllendorf). Berlin: Springer 1927.
Lagally, H.: Beiträge zur normalen und pathologischen Histologie des Labyrinthes. (Hauskatze.) Beitr. Anat. etc. Ohr. 5, 73—90 (1912).
Leidler, R.: Experimentelle Untersuchungen über das Endingungsgebiet des Nervus vestibularis. Arb. neurol. Inst. Univ. Wien 21, 151—212 (1914).
Lenhossék, M. v.: Die Nervenendigungen in den Maculae und Cristae acusticae. In: Beiträge zur Histologie des Nervensystems und der Sinnesorgane, p. 1—37 (ed. M. v. Lenhossék). Wiesbaden 1894.
Lindeman, H. H.: Cellular pattern and nerve supply of the vestibular sensory epithelia. Acta oto-laryng. (Stockh.), Suppl. 224, 86—95 (1967).
— Regional differences in sensitivity of the vestibular sensory epithelia to ototoxic antibiotics. Acta oto-laryng. (Stockh.). 67, 177—189 (1969).
Lorente de Nó, R.: Études sur l'anatomie et la physiologie du labyrinthe de l'oreille et du VIIIe nerf. Deuxième partie. Quelques données au sujet de l'anatomie des organes sensoriels du labyrinthe. Trav. Lab. Rech. Biol. Univ. Madr. 24, 53—153 (1926).
— Ausgewählte Kapitel aus der vergleichenden Physiologie des Labyrinthes. Die Augenmuskelreflexe beim Kaninchen und ihre Grundlagen. Ergebn. Physiol. 32, 73—242 (1931).
— Anatomy of the eighth nerve. The central projection of the nerve endings of the internal ear. Laryngoscope (St. Louis) 43, 1—38 (1933).
Lowenstein, O.: The functional significance of the ultrastructure of the vestibular end organs. In: The role of the vestibular organs in space exploration, p. 73—87. NASA SP-115. National Aeronautics and Space Administration, Washington, D.C. 1966.
— Functional aspects of vestibular structure. In: Myotatic, kinesthetic and vestibular mechanisms, p. 121—128. The Ciba Foundation Symp. London: Churchill 1967.
—, and M. P. Osborne: Ultrastructure of the sensory hair cells in the labyrinth of the ammocaeta larva of the lamprey, Lampetra fluviatilis. Nature (Lond.) 204, 197—198 (1964).
— —, and J. Wersäll: Structure and innervation of the sensory epithelia of the labyrinth in the thornback ray (Raja clavata). Proc. roy. Soc. Med. (Lond.) 160, 1—12 (1964).
—, and T. D. M. Roberts: The equilibrium function of the otolith organs of the thornback ray (Raja clavata). J. Physiol. (Lond.) 110, 392—415 (1950).
—, and A. Sand: The mechanism of the semicircular canal. A study of the responses of single fibre preparations to angular accelerations and to rotation at constant speed. Proc. roy. Soc. B 129, 256—275 (1940).
—, and J. Wersäll: A functional interpretation of the electron-microscopic structure of the sensory hairs in the cristae of the elasmobranch Raja clavata in terms of directional sensitivity. Nature (Lond.) 184, 1807 (1959).
Lundquist, P.-G.: The endolymphatic duct and sac in the guinea pig. Acta oto-laryng. (Stockh.), Suppl. 201, 1—108 (1965).
Maillet, M.: Le réactif au tétraoxyde d'osmium-iodure de zinc. Rev. méd. Tours 4, 247—268 (1963).
Mair, I. W. S., and C. Fernández: Pathological and functional changes following hemisection of the lateral ampullary nerve. Acta oto-laryng. (Stockh.) 62, 513—531 (1966).
Neubert, K.: Die Basilarmembran des Menschen und ihr Verankerungssystem. Z. Anat. Entwickl.-Gesch. 114, 539—588 (1950).
— Zur morphologischen Erfassung der Ansprechgebiete im Innenohr. Verh. anat. Ges. 50, 204—209 (1952).
Neumann, G., u. K. Neubert: Die Sensularien des Innenohres unter der Einwirkung von Streptomycin. Arzneimittel-Forsch. 8, 63—72 (1958).
Niemack, J.: Maculae und Cristae acusticae mit Ehrlich's Methylenblaumethode. Anat. H. 2, 205—234 (1892/93).

Nomura, Y., R. R. Gacek, and K. Balogh: Efferent innervation of vestibular labyrinth: Histochemical demonstration of acetylcholinesterase activity in guinea pig inner ear. Arch. Otolaryng. 81, 335—339 (1965).

Odenius, M. V.: Ueber das Epithel der Maculae acusticae beim Menschen. Arch. mikr. Anat. 3, 115—135 (1867).

Oort, H.: Über die Verästelung des Nervus octavus bei Säugetieren. Anat. Anz. 51, 272—280 (1918/19).

Palumbi, G.: The innervation of the human inner ear in the light of new histomorphological studies. Sci. med. ital. 3, 351—367 (1954).

Perlman, H. B.: The saccule. Observations on a differentiated reinforced area of the saccular wall in man. Arch. Otolaryng. 32, 678—691 (1940).

Petroff, A. E.: An experimental investigation of the origin of efferent fibre projections to the vestibular neuroepithelium. Anat. Rec. 121, 352—353 (1955).

Poljak, S.: Über die doppelte Innervation der Macula sacculi und über das cochleo-vestibuläre Bündel bei den Säugetieren. Z. Anat. Entwickl.-Gesch. 84, 144—152 (1927a).

— Über die Nervenendigungen in den vestibulären Sinnesendstellen bei den Säugetieren. Z. Anat. Entwickl.-Gesch. 84, 131—144 (1927b).

Pritchard, U.: The termination of the nerves in the vestibule and semicircular canals of mammals. Quart. J. micr. Sci. 16, 398—404 (1876).

Rasmussen, A. T.: Studies of the VIIIth cranial nerve of man. Laryngoscope (St. Louis) 50, 67—83 (1940).

Rasmussen, G. L.: The olivary peduncle and other fiber projections of the superior olivary complex. J. comp. Neurol. 84, 141—220 (1946).

— Further observations of the efferent cochlear bundle. J. comp. Neurol. 99, 61—74 (1953).

— Efferent fibers of the cochlear nerve and cochlear nucleus. In: Neural mechanisms of the auditory and vestibular systems, p. 105—115 (eds. G. L. Rasmussen and W. F. Windle). Springfield (Ill.): Thomas 1960.

— A method of staining the statoacoustic nerve in bulk with Sudan black B. Anat. Rec. 139, 465—469 (1961).

—, and R. Gacek: Concerning the question of an efferent fiber component of the vestibular nerve of the cat. Anat. Rec. 130, 361—362 (1958).

Rauch, S.: Entwicklung, Transport und Resorption der Innenohrflüssigkeiten. Endolymphe. In: Biochemie des Hörorgans, p. 278—315 (ed. S. Rauch). Stuttgart: Thieme 1964.

Retzius, G.: Om hörselsnervens ändningssätt i maculae och cristae acusticae. Nord. Med. ark. 3, 1—4 (1871).

— Das Gehörorgan der Wirbelthiere. I. Das Gehörorgan der Fische und Amphibien. Stockholm: Der Centraldruckerei 1881a.

— Ueber die peripherische Endigungsweise des Gehörnerven. Biol. Untersuch. (Stockh.) 51—60 (1881b).

— Das Gehörorgan der Wirbelthiere. II. Das Gehörorgan der Reptilien, der Vögel und der Säugethiere. Stockholm: Der Centraldruckerei 1884.

— Die Endigungsweise des Gehörnerven. Biologische Untersuchungen (Stockh.), N.F. 3, 29—36 (1892).

— Zur Entwicklung der Zellen des Ganglion spirale acustici und zur Endigungsweise des Gehörnerven bei den Säugetieren. Biol. Untersuch. (Stockh.), N.F. 6, 52—57 (1894).

— Ueber die Endigungsweise des Gehörnerven in den Maculae und Cristae acusticae im Gehörlabyrinth der Wirbeltiere. Eine historisch-kritische Uebersicht. Biol. Untersuch. (Stockh.), N.F. 12, 21—32 (1905).

Romeis, B.: Mikroskopische Technik. München: Leibnitz 1948.

Rossi, G.: L'acetylcholinesterase au cours du developpement de l'oreille interne du cobaye. Acta oto-laryng. (Stockh.), Suppl. 170, 1—91 (1961).

—, and G. Cortesina: Research on the efferent innervation of the inner ear. J. Laryng. 77, 202—233 (1963).

— — The "efferent cochlear and vestibular system" in Lepus cuniculus L. Acta Anat. (Basel) 60, 362—381 (1965a).

— — The efferent innervation of the inner ear. A historical-bibliographical survey. Laryngoscope (St. Louis) 75, 212—235 (1965b).

Rosenhall, U.: Manuscript in preparation.

Sala, O.: The efferent vestibular system. Electrophysiological research. Acta oto-laryng. (Stockh.), Suppl. 197, 1—34 (1965).

Sasaki, H., u. J. Miyata: Experimentelle Studien über Otolithen. Z. Laryng. Rhinol. 34, 740—748 (1955).

Saxén, A.: Histological studies on the endolymph secretion and resorption in the inner ear. Acta oto-laryng. (Stockh.) 40, 23—31 (1951).

Scarpa, A.: Anatomische Untersuchungen des Gehörs und Geruchs. Nürnberg 1800.

Schmidt, R. S.: Frog labyrinthine efferent impulses. Acta oto-laryng. (Stockh.) 56, 51—64 (1963).

Schultze, M.: Ueber die Endigungsweise des Hörnerven im Labyrinth. Arch. Anat. Physiol. wiss. Med. 343—381 (1858).

Shute, C. C. D.: The anatomy of the eighth cranial nerve in man. Proc. roy. Soc. Med. 44, 1013—1018 (1951).

Smith, C. A.: Microscopic structure of the utricle. Ann. Otol. (St. Louis) 65, 450—469 (1956).

— Discussion in: The role of the vestibular organs in the exploration of space, p. 381—383. NASA SP-77. National Aeronautics and Space Administration, Washington, D.C. 1965.

— Utricle and saccule. In: Submicroscopic structure of the inner ear, p. 175—195 (ed. S. Iurato). Oxford and London: Pergamon Press 1967.

Spoendlin, H. H.: Organization of the sensory hairs in the gravity receptors in utricle and saccule of the squirrel monkey. Z. Zellforsch. 62, 701—716 (1964).

— Ultrastructural studies of the labyrinth in squirrel monkeys. In: The role of the vestibular organs in the exploration of space, p. 7—22. NASA SP-77. National Aeronautics and Space Administration, Washington, D.C. 1965.

— Some morphofunctional and pathological aspects of the vestibular sensory epithelia. In: The role of the vestibular organs in space exploration, p. 99—115. NASA SP-115. National Aeronautics and Space Administration, Washington, D.C. 1966a.

— Ultrastructure of the vestibular sense organ. In: The vestibular system and its diseases, p. 39—68 (ed. R. J. Wolfson). Philadelphia: University of Pennsylvania Press 1966b.

— Zur Ototoxizität des Streptomyzins. Pract. oto-rhino-laryng. (Basel) 28, 305—322 (1966c).

—, and W. Lichtensteiger: The adrenergic innervation of the labyrinth. Acta oto-laryng. (Stockh.) 61, 423—434 (1966).

— H. F. Shuknecht, and A. Graybiel: The ultrastructure of the otolith organs in squirrel monkeys after exposure to high levels of gravito-inertial force. BuMed Project MR 005. 13-6001 Subtask 1, Report No. 102 and NASA Order No. R-93. Pensacola, Fla.: Naval School of Aviation Medicine, p. 1—23 (1964).

Steifensand, K.: Untersuchungen über die Ampullen des Gehörorgans. Arch. Anat. Physiol. wiss. Med. 171—189 (1835).

Stein, B. M., and M. B. Carpenter: Central projections of portion of the vestibular ganglia innervating specific parts of the labyrinth in the Rhesus monkey. Amer. J. Anat. 120, 281—317 (1967).

Steinhausen, W.: Über die Funktion der Cupula in den Bogengangsampullen des Labyrinthes. Z. Hals-, Nas.- u. Ohrenheilk. 34, 201—209 (1933).

Stricht, N. van der: L'histogenèse des parties constituantes du neuroépithélium acoustique, des taches et des crètes acoustiques et de l'organe de Corti. Arch. Biol. (Liège) 23, 541—693 (1908).

— The genesis and structure of the membrana tectoria and the crista spiralis of the cochlea. Contr. Embryol. Carneg. Instn 7, 57—86 (1918).

Trincker, D.: Bestandspotentiale im Bogengangssystem des Meerschweinchens und ihre Änderungen bei experimentellen Cupula-Ablenkungen. Pflügers Arch. ges. Physiol. 264, 351—382 (1957).

— Neuere Untersuchungen zur Elektrophysiologie des Vestibularapparates. Naturwissenschaften 46, 344—350 (1959).

— Physiologie des Gleichgewichtsorgans. In: Hals-Nasen-Ohrenheilkunde, p. 311—361 (eds. J. Berendes, R. Link und F. Zöllner). Stuttgart: Thieme 1965.

Vilstrup, T.: Studies on the structure and function of the semicircular canals. Copenhagen: Munksgaard 1950.

Voit, M.: Zur Frage der Verästelung des Nervus acusticus bei den Säugetieren. Anat. Anz. **31**, 635—640 (1907).

Walberg, F., D. Bowsher, and A. Brodal: The termination of primary vestibular fibers in the vestibular nuclei in the cat. An experimental study with silver methods. J. comp. Neurol. **110**, 391—419 (1958).

Werner, C. F.: Über Volumenveränderungen der Cupula terminalis im Ohrlabyrinth, insbesondere durch Fixation, Entkalkung und Einbettung. Z. Zellforsch. **16**, 471—483 (1932).

— Die Differenzierung der Maculae im Labyrinth, insbesondere bei Säugetieren. Z. Anat. Entwickl.-Gesch. **99**, 696—709 (1933).

— Funktionelle und vergleichende Anatomie der Otolithenapparate bei den Vögeln. Z. Anat. Entwickl.-Gesch. **108**, 775—791 (1938).

— Das Labyrinth. Leipzig. Thieme 1940.

— Das Gehörorgan der Wirbeltiere und des Menschen. Leipzig: Thieme 1960.

Wersäll, J.: The minute structure of the crista ampullaris in the guinea pig as revealed by the electron microscope. Acta oto-laryng. (Stockh.) **44**, 359—369 (1954).

— Studies on the structure and innervation of the sensory epithelium of the cristae ampullares in the guinea pig. A light and electron microscopic investigation. Acta oto-laryng. (Stockh.), Suppl. **126**, 1—85 (1956).

— Electron micrographic studies of vestibular hair cell innervation. In: Neural mechanisms of the auditory and vestibular systems, p. 247—257 (eds. G. L. Rasmussen and W. F. Windle). Springfield (Ill.): Thomas 1960.

— Vestibular receptor cells in fish and mammals. Acta oto-laryng. (Stockh.), Suppl. **163**, 25—29 (1961).

— Cristae ampullares. In: Submicroscopic structure of the inner ear, p. 195—210 (ed. S. Iurato). Oxford and London: Pergamon Press 1967.

— H. Engström, and S. Hjorth: Fine structure of the guinea pig macula utriculi. Acta oto-laryng. (Stockh.), Suppl. **116**, 298—303 (1954).

—, and Å. Flock: Functional anatomy of the vestibular and lateral line organs. In: Contributions to sensory physiology, p. 39—61 (ed. W. D. Neff). New York and London: Academic Press 1965.

— —, and P.-G. Lundquist: Structural basis for directional sensitivity in cochlear and vestibular sensory receptors. Cold Spr. Harb. Symp. quant. Biol. **30**, 115—145 (1966).

—, L. Gleisner, and P.-G. Lundquist: Ultrastructure of the vestibular end organs. In: Myotatic, kinesthetic and vestibular mechanisms, p. 105—116. The Ciba Foundation Symposium. London: J. & A. Churchill Limited 1967.

—, and P.-G. Lundquist: Morphological polarization of the mechanoreceptors of the vestibular and acoustic systems. In: The role of the vestibular organs in space exploration, p. 57—71. NASA SP-115. National Aeronautics and Space Administration, Washington, D.C. 1966.

Weston, J. K.: Observations on the distribution of ganglion cells and fibers related to the saccule and the basal coil of the cochlea. Acta neerl. Morph. **1**, 136—150 (1937).

— Observations on the comparative anatomy of the VIIIth nerve complex. Acta oto-laryng. (Stockh.) **27**, 457—497 (1939).

Winther, F. Ö.: Early degenerative changes in the inner ear sensory cells of the guinea pig following local x-ray irradiation. A preliminary report. Acta oto-laryng. (Stockh.) **67**, 262—268 (1969).

Wislocki, G. B., and A. J. Ladman: Selective and histochemical staining of the otolithic membrane of the inner ear. J. Anat. (Lond.) **89**, 3—12 (1955).

Wolff, D., R. J. Bellucci, and A. A. Eggston: Microscopic anatomy of the temporal bone. Baltimore: Williams & Wilkins Co. 1957.

Subject Index